Advances in Applied Mathematical Analysis and Applications

RIVER PUBLISHERS SERIES IN MATHEMATICAL AND ENGINEERING SCIENCES

Series Editors:

TADASHI DOHI
Hiroshima University
Japan

ALIAKBAR MONTAZER HAGHIGHI
Prairie View Texas A&M University
USA

MANGEY RAM
Graphic Era University
India

Indexing: all books published in this series are submitted to the Web of Science Book Citation Index (BkCI), to SCOPUS, to CrossRef and to Google Scholar for evaluation and indexing.

Mathematics is the basis of all disciplines in science and engineering. Especially applied mathematics has become complementary to every branch of engineering sciences. The purpose of this book series is to present novel results in emerging research topics on engineering sciences, as well as to summarize existing research. It engrosses mathematicians, statisticians, scientists and engineers in a comprehensive range of research fields with different objectives and skills, such as differential equations, finite element method, algorithms, discrete mathematics, numerical simulation, machine leaning, probability and statistics, fuzzy theory, etc.

Books published in the series include professional research monographs, edited volumes, conference proceedings, handbooks and textbooks, which provide new insights for researchers, specialists in industry, and graduate students.

Topics covered in the series include, but are not limited to:

- Advanced mechatronics and robotics
- Artificial intelligence
- Automotive systems
- Discrete mathematics and computation
- Fault diagnosis and fault tolerance
- Finite element methods
- Fuzzy and possibility theory
- Industrial automation, process control and networked control systems
- Intelligent control systems
- Neural computing and machine learning
- Operations research and management science
- Optimization and algorithms
- Queueing systems
- Reliability, maintenance and safety for complex systems
- Resilience
- Stochastic modelling and statistical inference
- Supply chain management
- System engineering, control and monitoring
- Tele robotics, human computer interaction, human-robot interaction

For a list of other books in this series, visit www.riverpublishers.com

Advances in Applied Mathematical Analysis and Applications

Editors

Mangey Ram

Graphic Era Deemed to be University
India

Tadashi Dohi

Hiroshima University
Japan

LONDON AND NEW YORK

Published 2019 by River Publishers
River Publishers
Alsbjergvej 10, 9260 Gistrup, Denmark
www.riverpublishers.com

Distributed exclusively by Routledge
4 Park Square, Milton Park, Abingdon, Oxon OX14 4RN
605 Third Avenue, New York, NY 10017, USA

First issued in paperback 2023

Advances in Applied Mathematical Analysis and Applications / by Mangey Ram, Tadashi Dohi.

Routledge is an imprint of the Taylor & Francis Group, an informa business

Publisher's Note
The publisher has gone to great lengths to ensure the quality of this reprint but points out that some imperfections in the original copies may be apparent.

While every effort is made to provide dependable information, the publisher, authors, and editors cannot be held responsible for any errors or omissions.

ISBN 13: 978-87-7022-970-8 (pbk)
ISBN 13: 978-87-7022-110-8 (hbk)
ISBN 13: 978-1-003-33707-2 (ebk)

Contents

3   Multiparametric Modeling of Carbon Cycle in Temperate Wetlands for Regional Climate Change Analysis Using Satellite Data   **51**

Anna Kozlova, Lesia Elistratova, Yuriy V. Kostyuchenko, Alexandr Apostolov and Igor Artemenko

Preface

In recent years, applied mathematics has been used in all novel disciplines of scientific evolution and development. The book **Advances in Applied Mathematical Analysis and Applications** is published with the purpose to summarize the interdisciplinary work of applied mathematics.

Chapter 1 discusses the method of similarity solutions of spherical shock waves in a self-gravitating ideal gas. This method is generally used to study the continuous symmetry in mathematics, mechanics, and theoretical physics, and helps us to simplify the complicated problem involving physical phenomena into solvable mathematical system of equations.

In Chapter 2, the influences of the magnetic field and the Navier's slip condition on the boundary layer flow of Cu – water nanofluid nearby a stagnation point over a stretching surface are discussed.

Chapter 3 discusses the advantages of applied mathematics-based approaches for ecology, environmental security, climate regional studies oriented to change adaptation, where the remote sensing technologies are demonstrated. Different approaches to anthropogenic and natural emissions inventory and analysis are also described and discussed.

Chapter 4 describes a methodology for developing a smart system by utilising the application of an adaptive neuro-fuzzy inference system that can be used by physicians to accelerate the diagnosis process. This aims at developing a capable intelligent system that can be used for the classification purpose.

Chapter 5 considers an inventory model where demand and deterioration rates are formulated in fuzzy environment, because, in real-life situations, demand and deterioration rates of an item are slightly disrupted from their original values.

In Chapter 6, the authors give an introductory discussion on summability process; regularity of summability process and its Silverman–Toeplitz theorem. Further, they explain about the stability of the frequency response of the system.

Chapter 7 concerns the various parameters that are considered in the design of the manufacturing control and automation. Manufacturing system can be designed in accordance with the techniques used to provide the final state of products that may be finished or semi-finished. Additive, subtractive, and rapid prototyping is proposed as a technique designed for the manufacturing which can be used to produce various types of products according to their complexity and requirement of the production procedure.

In Chapter 8, an application of SEIR model is studied for the crop through water and soil texture. The resulting study reflects the behaviour of the crop yield that occurs when crop comes in the contact of other model parameters.

Chapter 9 briefly describes the importance of mesh-free techniques using radial basis function approach in solving various complex phenomena in the framework of the differential equation.

In Chapter 10, the author applies the Rayleigh–Ritz method to study the time-period of a non-homogeneous square plate with linear variation in thickness and temperature, on clamped edge conditions. A comparison of frequency modes with the existing results in the literature is also given.

Chapter 11 provides a brief historical survey of the results, definitions, mathematical lemmas, and theorems, in complete metric spaces on a fixed point for maps concerning several integral types of contractive conditions.

In Chapter 12, the variations of stripping factors for various energies of the gamma rays in airborne surveys are studied from the viewpoint of direct measurements for radiometric counts in different energy windows.

Chapter 13 deals with the inverting of the area temperature profile $a(T)$ from the measured total power spectrum $W(v)$ of temperature at different frequencies. The inversion solution of this black body radiation problem is derived using the Planck radiation law.

The editors would like to thank all the authors for their great contributions to this book as well as all anonymous referees for their excellent jobs in reviewing the articles. We hope that the readers can lean the latest results and trends in respective research fields and will be inspired to make their further research and development in applied mathematics.

Mangey Ram
India

Tadashi Dohi
Japan

Acknowledgements

The editors acknowledge River Publishers for this opportunity and professional support. Also, they would like to thank all the chapter authors and reviewers for their availability for this work.

List of Contributors

Abdu Masanawa Sagir, *Department of Basic Studies, Hassan Usman Katsina Polytechnic, Katsina, Nigeria; E-mail: amsagir@yahoo.com*

Aeshwarya Dixit, *University School of Business, Chandigarh University, Gharuan, Punjab-140301, India; E-mail: aeshwaryagarg1017@gmail.com*

Alexandr Apostolov, *Scientific Centre for Aerospace Research of the Earth, National Academy of Sciences of Ukraine, Kyiv, Ukraine; E-mail: alex@casre.kiev.ua*

Alka Munjal, *National Institute of Technology Kurukshetra, Haryana, India; E-mail: alkamunjal8@gmail.com*

Amit Sharma, *Department of Mathematics, Amity University Haryana, India; E-mail: dba.amitsharma@gmail.com*

Anna Kozlova, *Scientific Centre for Aerospace Research of the Earth, National Academy of Sciences of Ukraine, Kyiv, Ukraine; E-mail: ak.koann@gmail.com*

Astha Chauhan, *Department of Applied Science and Engineering, Indian Institute of Technology Roorkee, India; E-mail: asthaiitr1@gmail.com*

Ekta N. Jayswal, *Department of Mathematics, Gujarat University, Ahmedabad, 380009, Gujarat, India; E-mail: jayswal.ekta1993@gmail.com*

Foram A. Thakkar, *Department of Mathematics, Gujarat University, Ahmedabad, 380009, Gujarat, India; E-mail: foramkhakhar3@gmail.com*

Geeta Arora, *Lovely Professional University, Phagwara, Punjab, India; E-mail: geetadma@gmail.com*

Gurpreet Singh Bhatia, *Lovely Professional University, Phagwara, Punjab, India; E-mail: gurpreetsidakbhatia@gmail.com*

Igor Artemenko, *Scientific Centre for Aerospace Research of the Earth, National Academy of Sciences of Ukraine, Kyiv, Ukraine; E-mail: igor.artemenko@casre.kiev.ua*

KM Kanika, *Department of Mathematics, Malaviya National Institute of Technology, Jaipur-302017, India; E-mail: kanikatomar94@gmail.com*

Lesia Elistratova, *Scientific Centre for Aerospace Research of the Earth, National Academy of Sciences of Ukraine, Kyiv, Ukraine; E-mail: tkach_lesya@ukr.net*

Magfura Pervin, *Department of Applied Mathematics with Oceanology and Computer Programming Vidyasagar University, Midnapore-721102, West Bengal, India; E-mail: mamoni2014@rediffmail.com*

Mohit Pant, *Department of Mechanical Engineering, National Institute of Technology, Hamirpur, Himachal Pardesh-177001, India; E-mail: mohitpant.iitr@gmail.com*

Mohit Sood, *Department of Mechanical Engineering, Rayat and Bahara Institute of Engineering and Nano Technology, Hoshiarpur, Punjab-146104, India; E-mail: msood04@gmail.com*

Moksha H. Satia, *Department of Mathematics, Gujarat University, Ahmedabad, 380009, Gujarat, India; E-mail: mokshasatia.05@gmail.com*

Naveen Mani, *Department of Mathematics, Sandip University, Nashik, Maharashtra, India; E-mail: naveenmani81@gmail.com*

Nita H. Shah, *Department of Mathematics, Gujarat University, Ahmedabad, 380009, Gujarat, India; E-mail: nitahshah@gmail.com*

Rajan Arora, *Department of Applied Science and Engineering, Indian Institute of Technology Roorkee, India; E-mail: rajanfpt@iitr.ac.in*

Reeta Bhardwaj, *Department of Mathematics, Amity University Haryana, India; E-mail: bhardwajreeta84@gmail.com*

Sahil Garg, *Department of Mechanical Engineering, National Institute of Technology, Hamirpur, Himachal Pardesh-177001, India; E-mail: sahil.garg1017@gmail.com*

Sankar Kumar Roy, *Department of Applied Mathematics with Oceanology and Computer Programming, Vidyasagar University, Midnapore-721102, West Bengal, India; E-mail: sankroy2006@gmail.com*

Santosh Chaudhary, *Department of Mathematics, Malaviya National Institute of Technology, Jaipur-302017, India; E-mail: d11.santosh@yahoo.com*

Saratha Sathasivam, *School of Mathematical Sciences, Universiti Sains Malaysia, 11800 Pulau Pinang, Malaysia; E-mail: saratha@usm.my*

Satishkumar M. Bhati, *Department of Mathematics, Sandip University, Nashik, Maharashtra, India; E-mail: satishmbhati@gmail.com*

Shivvaran Singh Raghuwanshi, 1. *Ex-Head Physics Group, Atomic Minerals Directorate for Exploration & Research, Department of Atomic Energy, Hyderabad, India*
2. *Bhopal (M.P.), India; E-mail: raghu2257@yahoo.com*

Smita Sonker, *National Institute of Technology Kurukshetra, Haryana, India; E-mail: smita.sonker@gmail.com*

Yuriy V. Kostyuchenko, *Scientific Centre for Aerospace Research of the Earth, National Academy of Sciences of Ukraine, Kyiv, Ukraine; E-mail: yuriy.v.kostyuchenko@gmail.com*

List of Figures

List of Tables

List of Abbreviations

AGRS	Airborne Gamma-Ray Spectometric
AIRS	Atmospheric Infrared Sounder
AMD	Atomic Minerals Directorate for Exploration and Research
AMSU	Advanced Microwave Sounding Unit
CCSM	Community Climate System Model
CESM	Community Earth System Model
CLM	Climate Limited-area Model
DAO	Data Assimilation Office
ENVISAT	ENVIronmental SATellite
EOS	Earth Observation Satellite
ETKF	Ensemble Transform Kalman Filter
FPAR	Fraction of Absorbed Photosynthetically Active Radiation
GHG	Greenhouse Gases
GOSAT	the Greenhouse Gases Observing Satellite
GPP	Gross Primary Production
IPCC	the Intergovernmental Panel on Climate Change
KPCA	Kernel Principal Component Analysis
MODIS	The Moderate Resolution Imaging Spectroradiometer
NDVI	Normalized Difference Vegetation Index
NMVOC	Non-Methane Volatile Organic Compounds
NPP	Net Primary Production
OLS	Ordinary Least Squares
SCIAMACHY	SCanning Imaging Absorption SpectroMeter for Atmospheric CHartographY
WFM-DOAS	Weighting Function Modified Differential Optical Absorption Spectroscopy
WLS	Weighted Least Squares

1

Similarity Solutions of Spherical Shock Waves in a Self-Gravitating Ideal Gas

Astha Chauhan[*] and Rajan Arora

Department of Applied Science and Engineering, Indian Institute
of Technology Roorkee, India
E-mail: asthaiitr1@gmail.com; rajanfpt@iitr.ac.in
*Corresponding Author

Lie group of transformations method is used to obtain the similarity solutions
for the system of non-linear partial differential equations, which describes
the one-dimensional unsteady adiabatic flow of a self-gravitating ideal gas
with the effect of magnetic field. The density of the medium is supposed to
be varying and obeying the exponential law. Two cases have been discussed
for the possible solutions for different values of arbitrary constants appearing
in the set of infinitesimal generators. With the help of infinitesimal generators,
the given system of non-linear PDEs is converted into a system of ODEs.
In this work, a particular case of the collapse of an imploding spherical
shock is shown in detail. The effects of adiabatic index and the strength of
magnetic field are shown in detail. To obtain the profile of flow variables
and similarity exponent delta, numerical calculations have been performed
for different values of the ambient density exponent. All computational work
has been performed using the software Mathematica.

1.1 Introduction

Most of the physical phenomena in our world are essentially non-linear
and can be described by the mathematical models in the form of partial
differential equations. These PDEs play an important role in many areas
of physics: astrophysics, gas-dynamics, combustion theory, cosmology, etc.

With the help of the basic laws of mechanics and thermodynamics, the theoretical foundation of gas-dynamics is formed. To study the propagation of shock waves is important, because this phenomenon is well-marked in the disassembled nature of the absorption of energy. Shock waves form when the velocity of object becomes higher than the velocity of sound. Shocks waves are observed at many places such as the snapping of belts, nature, laboratory, etc. Strong converging shocks are of interest in material synthesis, where the hardness, phase and other properties of a material can be changed because of shock compression.

The method of Lie group is generally used to study the continuous symmetry in mathematics, mechanics and theoretical physics. This method helps us simplify the complicated problem involving physical phenomena into solvable mathematical system of equations. The basic idea of Lie group method for solving physical problem, which has been formulated in the form of mathematical equations, can be found in the work presented by Logan [12], Sharma and Radha [17], Bluman and Kumei [4], Arora et al. [2] and Hydon [8]. Sharma and Arora [18] used this method in their work and obtained the entire class of similarity solutions and also studied imploding shock in an ideal gas. A theoretical study of imploding shock wave was first performed by Guderley [7]. In 2012, Jena [11] used Lie group method and obtained the similarity solutions in a plasma with axial magnetic field (θ-pinch). The method of Lie group was used to study the propagation of shock wave through a non-ideal gas obeying the equation of state used in [5], interaction of weak discontinuities in [1], solution of the system of equations describing the flow at the stellar surface in the work of Saxena and Jena [16], self-similar solutions of equations describing the flow in a gas with dust particles [10].

Ray [6] studied the propagation of shock wave in an exponential medium and found exact solutions. Singh and Vishwakarma [19] have obtained the self-similar solutions of a shock wave in the presence of radiation heat flux, heat conduction and gravitational field in the mixture of a non-ideal gas. Nath et al. [14] discussed the flow behind the shock in a self gravitating gas in a medium with varying exponential density. They have shown the effects of gravitational constant, variation in magnetic field and adiabatic exponent for adiabatic and isothermal flows in an ideal gas.

In general, it is very hard to determine the solution of the system of quasi-linear hyperbolic PDEs without any approximations. Hence, we seek similarity solutions subject to the shock conditions along a set of similarity curves along which the sytem of PDEs transforms into the system of ODEs.

For determining the similarity solutions, there are two methods to obtain the similarity exponent δ. In the first method, δ is obtained by either conservation laws or by dimensional considerations and in the second method, the similarity exponent is determined by integrating the reduced ODEs [22]. In this paper, we have considered the similarity solutions by using the second method.

In the present work, we have considered a system of hydrodynamic equations governing one-dimensional unsteady spherical symmetric motion in the presence of azimuthal magnetic field which describes the flow in a self-gravitating gas. We have applied the Lie group method to solve the system (1.1) and obtained infinitesimal generators for this system. With the help of these generators, the system of partial differential equations is reduced into the system of ordinary differential equations. The main advantage of this method is to reduce the number of independent variables by one. A particular case of the collapse of an imploding shock is shown in detail for spherically symmetric flow. The effect of an increase in the value of specific heats ratio constant and the strength of magnetic field are shown in detail. The flow patterns of variables just behind the shock are obtained by performing numerical computations using the software Mathematica.

1.2 Formulation of Problem

The fundamental equations of motion governing one-dimensional unsteady adiabatic flow of electrically conducting and self-gravitating ideal gas, with the effect of magnetic field are given as follows (see [13, 14, 20, 21, 23]):

$$\rho_t + v\rho_r + \rho v_r + 2\frac{\rho v}{r} = 0,$$

$$v_t + vv_r + \frac{1}{\rho}\left[p_r + \mu h h_r + \mu \frac{h^2}{r}\right] + \frac{Gm}{r^2} = 0,$$

$$h_t + vh_r + hv_r + \frac{hv}{r} = 0, \tag{1.1}$$

$$m_r - 4\pi\rho r^2 = 0,$$

$$p_t + vp_r - \frac{\gamma p}{\rho}(\rho_t + v\rho_r) = 0,$$

where p is the pressure, ρ is the density, v is the velocity, h is the magnetic field, γ is adiabatic index, μ is the magnetic permeability and m is the mass

which is contained in a unit sphere of radius r. The effects of heat conduction and viscosity are not considered and electrical conductivity of the gas is supposed to be infinite. The term $\frac{Gm}{r^2}$ in the second equation of system (1.1) represents the gravitational effect, where G is the gravitational constant.
As the behaviour of the medium is supposed to be ideal, therefore

$$p = \Gamma \rho T, \quad e = \frac{p}{(\gamma - 1)\rho}, \tag{1.2}$$

where Γ is the gas constant, e is the internal energy per unit mass and γ is the adiabatic index.

Let $r = \bar{R}(t)$ denotes the shock wave front motion propagating into a medium specified by

$$u_0 = 0, \quad \rho = \rho_c r^\theta \quad h_0 = h_c \quad \text{and} \quad m_0 = m_c, \tag{1.3}$$

where the suffix ρ_c, θ, h_c and m_c are the given reference constants with the associated medium. Therefore, the jump conditions across the shock front for the considered problem is defined as [13]:

$$\rho_1(V - u_1) = \rho_0 V,$$

$$p_1 + \frac{\mu h_1^2}{2} + \rho_1(V - u_1)^2 = p_0 + \frac{\mu h_0^2}{2} + \rho_0 V^2,$$

$$e_1 + \frac{p_1}{\rho_1} + \frac{\mu h_1^2}{\rho_1} + \frac{1}{2}(V - u_1)^2 = e_0 + \frac{p_0}{\rho_0} + \frac{\mu h_0^2}{\rho_0} + \frac{1}{2}V^2, \tag{1.4}$$

$$h_1(V - u_1) = h_0 V,$$

$$m_1 = m_0,$$

where $V = \frac{d\bar{R}}{dt}$ be the velocity of shock shock front. Since the shock is strong, therefore $p_0 \simeq 0$, and $e_0 \simeq 0$.

Then from Equation (1.4), the boundary conditions across the shock propagating as follows:

$$v = (1 - \beta)V,$$

$$\rho = \frac{\rho_0}{\beta},$$

$$p = \left[(1 - \beta) + \frac{1}{2 M_A^2}\left(1 - \frac{1}{\beta^2}\right)\right]V^2 \rho_0, \tag{1.5}$$

$$h = \frac{h_0}{\beta},$$

$$m = m_0,$$

where $M_A = \sqrt{\frac{\rho_0 V^2}{\mu h_0^2}}$, is the Alfven-Mach number. The value of $\beta\,(0 < \beta < 1)$ is determined by the expression given as follows:

$$\beta^2 - \beta\left(\frac{\gamma(M_A^{-2} + 1) - 1}{(\gamma + 1)}\right) + \frac{(\gamma - 2)M_A^{-2}}{(\gamma + 1)} = 0. \tag{1.6}$$

1.3 Similarity Analysis

For obtaining the similarity solutions of the system (1.1), we determine its symmetry group [15] such that (1.1) remains invariant under this symmetry group of transformations. So, we consider the infinitesimal group of transformations as follows:

$$
\begin{aligned}
r^* &= r + \epsilon\chi(\rho, v, p, h, m, r, t), \quad t^* = t + \epsilon\psi(\rho, v, p, h, m, r, t), \\
v^* &= v + \epsilon U(\rho, v, p, h, m, r, t), \quad \rho^* = \rho + \epsilon S(\rho, v, p, h, m, r, t), \quad (1.7) \\
p^* &= p + \epsilon P(\rho, v, p, h, m, r, t), \quad h^* = h + \epsilon E(\rho, v, p, h, m, r, t), \\
m^* &= m + \epsilon F(\rho, v, p, h, m, r, t),
\end{aligned}
$$

where χ, ψ, U, S, P, E and F are the infinitesimal generators such that the system (1.1) along with the shock conditions (1.5) remains invariant under the above group (1.7), ϵ is very small quantity such that we can neglect its second and higher power terms. This group of transformations decreases the number of independent variables of the given system by one and thus, we get the system of ordinary differential equations.

We introduce the notation as

$$x_1 = t, \; x_2 = r, \; u_1 = \rho, \; u_2 = v, \; u_3 = p, \; u_4 = h, \; u_5 = m,$$

and

$$p_j^i = \frac{\partial u_i}{\partial x_j}, \quad \text{where } i = 1, 2, 3, 4, 5 \text{ and } j = 1, 2.$$

The system (1.1) of basic equations, which can be expressed as

$$F_s(x_j, u_i, p_j^i) = 0, \quad s = 1, 2, 3, 4, 5$$

is said to be invariant under the Lie group of transformations (1.7) if

$$LF_s = \alpha_{sr} F_r, \quad \text{where } s = 1, 2, 3, 4, 5 \text{ and } r = 1, 2, 3, 4, 5 \tag{1.8}$$

where L represents the Lie derivative and is given as

$$L = \xi_x^j \frac{\partial}{\partial x_j} + \xi_u^i \frac{\partial}{\partial u_i} + \xi_{p_j}^i \frac{\partial}{\partial p_j^i},$$

with $\xi_x^1 = \psi$, $\xi_x^2 = \chi$, $\xi_u^1 = S$, $\xi_u^2 = U$, $\xi_u^3 = P$, $\xi_u^4 = E$, $\xi_u^5 = F$ and

$$\xi_{p_j}^i = \frac{\partial \xi_u^i}{\partial x_j} + \frac{\partial \xi_u^i}{\partial u_k} p_j^k - \frac{\partial \xi_x^l}{\partial x_j} p_l^i - \frac{\partial \xi_x^l}{\partial u_n} p_l^i p_j^n, \qquad (1.9)$$

where $l = 1, 2$, $j = 1, 2$, $n = 1, 2, 3, 4, 5$ and $k = 1, 2, 3, 4, 5$.

So, the Equation (1.8) can be written as

$$\xi_x^j \frac{\partial F_s}{\partial x_j} + \xi_u^i \frac{\partial F_s}{\partial u_i} + \xi_{p_j}^i \frac{\partial F_s}{\partial p_j^i} = \alpha_{sr} F_r, \quad \text{where } r, s = 1, 2, 3, 4, 5. \quad (1.10)$$

Substitution of $\xi_{p_j}^i$ from Equation (1.9) into Equation (1.10) yields a polynomial equation in p_j^i. Equating to zero the coefficients of p_j^i and $p_j^i p_l^n$ on both sides of Equation (1.10), we obtain a system of first-order PDEs in terms of the infinitesimals generators χ, ψ, U, S, P, E and F. Therefore, from the invariance of the first equation of the system (1.1), we obtain the system of determining equations given as follows:

$$S_\rho - \psi_t - v\,\psi_r = \alpha_{11} - \frac{\gamma\,p}{\rho}\,\alpha_{15},$$

$$S_v - \rho\,\psi_r = \alpha_{12}, \qquad S_p = \alpha_{15}, \qquad S_h = \alpha_{13}, \qquad S_m = 0,$$

$$U - \chi_t + v\,S_\rho - v\,\chi_r + \rho\,U_\rho = v\alpha_{11} - \frac{\gamma\,p\,v}{\rho}\alpha_{15},$$

$$S + v S_v + \rho\,U_v - \rho\,\chi_r = \rho\,\alpha_{11} + v\,\alpha_{12} + h\,\alpha_{13}, \qquad (1.11)$$

$$v\,S_p + \rho\,U_p = \frac{1}{\rho}\,\alpha_{12} + v\,\alpha_{15},$$

$$v\,S_h + \rho\,U_h = \frac{\mu\,h}{\rho}\,\alpha_{12} + v\,\alpha_{13},$$

$$v\,S_m + \rho\,U_m = \alpha_{14},$$

$$S_t + v\,S_r - \frac{2\rho\,v}{r^2}\chi + \frac{v}{r}2S + \frac{\rho}{r}2U = \frac{2\rho\,v}{r}\alpha_{11} + \left(\frac{\mu\,h^2}{\rho\,r} + \frac{Gm}{r^2}\right)\alpha_{12}$$

$$+ \frac{h\,v}{r}\alpha_{13} - 4\pi\,\rho\,r^2\,\alpha_{14}.$$

The invariance of the second equation of the system (1.1) yields

$$U_\rho = \alpha_{21} - \frac{\gamma\,p}{\rho}\,\alpha_{25}, \qquad U_v - v\,\psi_r - \psi_t = \alpha_{22},$$

$$U_p - \frac{1}{\rho}\,\psi_r = \alpha_{25}, \qquad U_h - \frac{\mu\,h}{\rho}\,\psi_r = \alpha_{23}, \qquad U_m = 0,$$

$$v\,U_\rho + \frac{1}{\rho}\,P_\rho + \frac{\mu\,h}{\rho}\,E_\rho = v\,\alpha_{21} - \frac{\gamma\,p\,v}{\rho}\,\alpha_{25},\tag{1.12}$$

$$U - \chi_t + v\,U_v + \frac{1}{\rho}\,P_v + \frac{\mu\,h}{\rho}\,E_v = \rho\,\alpha_{21} + v\,\alpha_{22} + h\,\alpha_{23},$$

$$-\frac{S}{\rho^2} + v\,U_p + \frac{1}{\rho}\,P_p - \frac{1}{\rho}\,\chi_r + \frac{\mu\,h}{\rho}\,E_p = \frac{1}{\rho}\,\alpha_{22} + v\,\alpha_{25},$$

$$\frac{\mu\,E}{\rho} - \frac{\mu\,h}{\rho^2}\,S + v\,U_h + \frac{1}{\rho}\,P_h + \frac{\mu\,h}{\rho}\,E_h - \frac{\mu\,h}{\rho}\,\chi_r = \frac{\mu\,h}{\rho}\,\alpha_{22} + v\,\alpha_{23},$$

$$v U_m + \frac{1}{\rho}\,P_m + \frac{\mu\,h}{\rho}\,E_m = \alpha_{24},$$

$$U_t + v\,U_r + \frac{1}{\rho}\,P_r + \frac{\mu\,h}{\rho}\,E_r - \frac{\mu\,h^2}{r\rho^2}\,S + \frac{2\mu\,h}{\rho\,r}\,E - \left(\frac{\mu\,h^2}{r^2} + \frac{2\,Gm}{r^3}\right)\chi$$

$$+ \frac{FG}{r^2} = \frac{2\rho\,v}{r}\,\alpha_{21} + \left(\frac{\mu\,h^2}{\rho\,r} + \frac{Gm}{r^2}\right)\alpha_{22} + \frac{h\,v}{r}\,\alpha_{23} - 4\pi\,r^2\,\alpha_{24}.$$

The invariance of the third equation of the system (1.1) yields

$$E_\rho = \alpha_{31} - \frac{\gamma\,p}{\rho}\,\alpha_{35}, \qquad E_u - h\,\psi_r = \alpha_{32}, \qquad E_p = \alpha_{35},$$

$$E_h - \psi_t - v\,\psi_r = \alpha_{33}, \qquad E_m = 0,$$

$$h\,U_\rho + v\,E_\rho = v\,\alpha_{31} - \frac{\gamma\,p\,v}{\rho}\,\alpha_{35},\tag{1.13}$$

$$E + h\,U_v - h\,\chi_r + v\,E_v = \rho\,\alpha_{31} + v\,\alpha_{32} + h\,\alpha_{33},$$

$$h\,U_p + v\,E_p = \frac{1}{\rho}\,\alpha_{32} + v\,\alpha_{35},$$

$$U - \chi_t + h\,U_h + v\,E_h - v\,\chi_r = \frac{\mu\,h}{\rho}\,\alpha_{32} + v\,\alpha_{33},$$

$$h\,U_m + v\,E_m = \alpha_{34},$$

$$\frac{U\,h}{r} - \frac{h\,v}{r^2}\,\chi + E_t + h\,U_r + v\,E_r + \frac{v}{r}\,E = \frac{2\rho\,v}{r}\,\alpha_{31} + \left(\frac{\mu\,h^2}{\rho\,r} + \frac{Gm}{r^2}\right)\alpha_{32}$$

$$+ \frac{h\,v}{r}\,\alpha_{33} - 4\pi\,\rho\,r^2\,\alpha_{34}.$$

The invariance of the fourth equation of the system (1.1) yields

$$\alpha_{42} = 0, \qquad \alpha_{43} = 0, \qquad \alpha_{45} = 0, \qquad \psi_r = 0, \qquad -\frac{\gamma\,p}{\rho}\,\alpha_{45} + \alpha_{41} = 0,$$

$$F_\rho = v\,\alpha_{41} - \frac{\gamma\,p\,v}{\rho}\,\alpha_{45}, \qquad F_v = \rho\,\alpha_{41} + v\,\alpha_{42} + h\,\alpha_{43},$$

$$F_p = \frac{1}{\rho}\,\alpha_{42} + v\,\alpha_{45}, \qquad F_h = \frac{\mu\,h}{\rho}\,\alpha_{42} + v\,\alpha_{43}, \qquad F_m - \chi_r = \alpha_{44}, \qquad (1.14)$$

$$F_r - 4\pi\,r^2\,S - 8\pi\,\rho\,r\,\chi = \frac{2\rho\,v}{r}\,\alpha_{41} + \left(\frac{\mu\,h^2}{\rho\,r} + \frac{Gm}{r^2}\right)\alpha_{42} + \frac{h\,v}{r}\,\alpha_{43} - 4\pi\,\rho\,r^2\,\alpha_{44}.$$

Finally, the invariance of the last equation of the system (1.1) yields

$$S\frac{\gamma\,p}{\rho^2} - \frac{\gamma\,p}{\rho}\,S_\rho + P_\rho - \frac{\gamma\,P}{\rho} + \frac{\gamma\,p}{\rho}\,\psi_t + \frac{\gamma\,p\,v}{\rho}\,\psi_r = \alpha_{51} - \frac{\gamma\,p}{\rho}\,\alpha_{55},$$

$$-\frac{\gamma\,P\,v}{\rho} - \frac{\gamma\,p\,U}{\rho} + S\frac{\gamma\,p\,v}{\rho^2} + \frac{\gamma\,p\,v}{\rho}\,\chi_r - \frac{\gamma\,p\,v}{\rho}\,S_\rho + \frac{\gamma\,p}{\rho}\,\chi_t + v\,P_\rho = v\,\alpha_{51} - \frac{\gamma\,p\,v}{\rho}\,\alpha_{55},$$

$$P_v - \frac{\gamma\,p}{\rho}\,S_v = \alpha_{52}, \qquad -\frac{\gamma\,p}{\rho}\,S_p + P_p - \chi_t - v\,\chi_r = \alpha_{55}, \qquad P_h - \frac{\gamma\,p}{\rho}\,S_h = \alpha_{53},$$

$$-\frac{\gamma\,p}{\rho}\,S_m + P_m = 0, \qquad v\,P_v - \frac{\gamma\,p\,v}{\rho}\,S_v = \rho\,\alpha_{51} + v\,\alpha_{52} + h\,\alpha_{53}, \qquad (1.15)$$

$$U - \chi_t - \frac{\gamma\,p\,v}{\rho}\,S_p + v\,P_p - v\,\chi_r = \frac{1}{\rho}\,\alpha_{52} + v\,\alpha_{55},$$

$$v\,P_h - \frac{\gamma\,p\,v}{\rho}\,S_h = \frac{\mu\,h}{\rho}\,\alpha_{52} + v\,\alpha_{53}, \qquad -\frac{\gamma\,p\,v}{\rho}\,S_m + v\,P_m = \alpha_{54},$$

$$v\,P_r - \frac{\gamma\,p}{\rho}\,S_t + P_t - \frac{\gamma\,p\,v}{\rho}\,S_r = \frac{2\rho\,v}{r}\,\alpha_{51} + \left(\frac{\mu\,h^2}{\rho\,r} + \frac{Gm}{r^2}\right)\alpha_{52} + \frac{h\,v}{r}\,\alpha_{53} - 4\pi\,\rho\,r^2\,\alpha_{54}.$$

Now, solving the above systems (1.11)–(1.15) of the determining equations, we obtain the following infinitesimal generators:

$$\chi = (\alpha_{22} + 2a)\,r, \qquad\qquad \psi = a\,t + b,$$
$$S = (\alpha_{11} + a)\,\rho, \qquad\qquad U = (\alpha_{22} + a)\,v,$$
$$P = (\alpha_{11} + 2\alpha_{22} + 3a)\,p, \qquad E = \frac{1}{2}(\alpha_{11} + 2\alpha_{22} + 3a)\,h, \qquad (1.16)$$
$$F = (3\alpha_{22} + 4a)\,m,$$

where α_{11}, α_{22}, a and b are the arbitrary constants.

1.4 Similarity Solutions

The arbitrary constants, appearing in the expressions (1.16), give rise to two cases of possible solutions.

Case 1: When $a \neq 0$ and $(\alpha_{22} + 2a) \neq 0$, we define the new variables $\bar{r}$ and $\bar{t}$ as

$$\bar{r} = r, \qquad \bar{t} = t + \frac{b}{a}. \qquad (1.17)$$

We find that the Equations (1.1)–(1.15) remain invariant under the above transformations (1.17). Therefore, a new set of infinitesimals in terms of new variables $\bar{r}$ and $\bar{t}$, is obtained as

$$\chi = (\alpha_{22} + 2a)\, r, \qquad\qquad \psi = at,$$
$$S = (\alpha_{11} + a)\, \rho, \qquad\qquad V = (\alpha_{22} + a)\, v,$$
$$P = (\alpha_{11} + 2\alpha_{22} + 3a)\, p, \quad E = \frac{1}{2}(\alpha_{11} + 2\alpha_{22} + 3a)\, h, \qquad (1.18)$$
$$F = (3\alpha_{22} + 4a)\, m.$$

The invariant surface conditions given in [12] are as follows:

$$\psi\, \rho_t + \chi\, \rho_r = S, \qquad \psi\, v_t + \chi\, v_r = U, \qquad \psi\, p_t + \chi\, p_r = P,$$
$$\psi\, h_t + \chi\, h_r = E, \qquad \psi\, m_t + \chi\, m_r = F. \qquad (1.19)$$

The equations in (1.18) together with Equation (1.19), after performing integration, yield the following forms of the flow variables:

$$\rho = t^{\left(1 + \frac{\alpha_{11}}{a}\right)} \widehat{S}(\xi), \qquad\qquad v = t^{(\delta - 1)} \widehat{U}(\xi),$$
$$p = t^{\left(2\delta - 1 + \frac{\alpha_{11}}{a}\right)} \widehat{P}(\xi), \qquad h = t^{\frac{1}{2}\left(2\delta - 1 + \frac{\alpha_{11}}{a}\right)} \widehat{E}(\xi), \qquad (1.20)$$
$$m = t^{(3\delta - 2)} \widehat{F}(\xi),$$

where

$$\delta = \frac{\alpha_{22} + 2a}{a}. \qquad (1.21)$$

The functions $\widehat{S}, \widehat{U}, \widehat{P}, \widehat{E}$ and $\widehat{F}$ depend only on the similarity variable ξ, which is determined as

$$\xi = \frac{r}{t^\delta}. \qquad (1.22)$$

Since the shock must be a similarity curve and at the shock, the value of ξ is constant. Hence, without any loss of generality the shock may be normalized at $\xi = 1$. Therefore, at $\xi = 1$, the expression for the shock path and the shock velocity V are given as

$$\bar{R}(t) = t^\delta, \qquad V = \frac{\delta \bar{R}}{t}. \qquad (1.23)$$

At the strong shock, $\xi = 1$, we have the following expressions for the functions ρ, v, p, h and m:

$$\rho|_{\xi=1} = t^{\left(1+\frac{\alpha_{11}}{a}\right)} \widehat{S}(1), \qquad\qquad v|_{\xi=1} = t^{(\delta-1)} \widehat{U}(1),$$

$$p|_{\xi=1} = t^{\left(2\delta-1+\frac{\alpha_{11}}{a}\right)} \widehat{P}(1), \qquad h|_{\xi=1} = t^{\frac{1}{2}\left(2\delta-1+\frac{\alpha_{11}}{a}\right)} \widehat{E}(1), \qquad (1.24)$$

$$m|_{\xi=1} = t^{(3\delta-2)} \widehat{F}(1).$$

From the invariance of jump conditions, we obtain the following forms of $\rho_0(r)$ and $m_0(r)$:

$$\rho_0(r) = \rho_c\, r^\theta, \quad m_0(r) = m_c\, r^\eta, \qquad (1.25)$$

and at the shock $\xi = 1$, the boundary conditions on the functions $\widehat{S}$, $\widehat{U}$, $\widehat{P}$, $\widehat{E}$, and $\widehat{F}$ are obtained as follows:

$$\widehat{U}(1) = (1 - \beta)\,\delta, \qquad\qquad\qquad \widehat{S}(1) = \frac{1}{\beta}\rho_c,$$

$$\widehat{P}(1) = \left[(1 - \beta) + \frac{1}{2M_A^2}\left(1 - \frac{1}{\beta^2}\right)\right]\delta^2\,\rho_c, \qquad \widehat{E}(1) = \frac{1}{M_A\beta}h_c,$$

$$\qquad (1.26)$$

$$\widehat{F}(1) = m_c,$$

together with

$$\theta = \frac{\alpha_{11} + a}{\alpha_{22} + 2a}, \qquad \eta = \frac{3\alpha_{22} + 4a}{\alpha_{22} + 2a}, \qquad (1.27)$$

Equation 1.26(iii) shows that for the existence of a similarity solution, it is necessary that M_A must be constant, which implies that δ and θ are not independent, but rather

$$\delta\theta + 2(\delta - 1) = 0. \qquad (1.28)$$

Using Equations (1.22), (1.23) and (1.25), we obtain the Equation (1.20) in the following forms:

$$\rho = \rho_0(\bar{R}(t))S^*(\xi), \qquad\qquad v = V\,U^*(\xi),$$
$$p = \rho_0(\bar{R}(t))V^2\,P^*(\xi), \qquad h = \rho_0^{1/2}V\,E^*(\xi), \qquad (1.29)$$
$$m = m_0(\bar{R}(t))\,F^*(\xi),$$

where $S^*(\xi) = \frac{\widehat{S}(\xi)}{\rho_c}$, $U^*(\xi) = \frac{\widehat{U}(\xi)}{\delta}$, $P^*(\xi) = \frac{\widehat{P}(\xi)}{\delta^2 \rho_c}$, $E^*(\xi) = \frac{\widehat{E}(\xi)}{\delta \rho_c^{1/2}}$ and $F^*(\xi) = \frac{\widehat{F}(\xi)}{m_c}$.

Substituting Equation (1.29) in the system (1.1) and using Equations (1.21), (1.23) and (1.29), we obtain the system of ordinary ODEs in terms of S^*, U^*, P^*, E^* and F^*, which on dropping the asterisk sign becomes

$$S\theta + (U - \xi)S' + SU' + \frac{2U}{\xi}S = 0,$$

$$(U - \xi)U'S + \frac{(\delta - 1)}{\delta}US + P' + \mu EE' + \frac{\mu E^2}{\xi} + \frac{Gm_c}{\xi^2 \delta^2}FS = 0,$$

$$(U - \xi)E' + \left[\frac{U}{\xi} + U'\right]E + \left[\frac{(\delta - 1)}{\delta} + \frac{\theta}{2}\right]E = 0, \qquad (1.30)$$

$$F' - \frac{4\pi \rho_c \xi^2}{m_c}S = 0,$$

$$(U - \xi)P' + \left[\frac{2(\delta - 1)}{\delta} + (1 - \gamma)\theta\right]P - \gamma P(U - \xi)\frac{S'}{S} = 0.$$

For the strong shocks, the R-H conditions are given by

$$U(1) = (1 - \beta), \qquad\qquad\qquad S(1) = \frac{1}{\beta},$$

$$P(1) = \left[(1 - \beta) + \frac{1}{2M_A^2}\left(1 - \frac{1}{\beta^2}\right)\right], \qquad E(1) = \frac{1}{M_A\beta}, \qquad (1.31)$$

$$F(1) = 1.$$

Case 2: Let us consider the case when $a = 0$ and $\alpha_{22} \neq 0$. Therefore, a new set of infinitesimals is obtained as

$$\begin{aligned}
\chi &= (\alpha_{22})\,r, & \psi &= b, \\
S &= (\alpha_{11})\,\rho, & V &= (\alpha_{22})\,v, \\
P &= (\alpha_{11} + 2\alpha_{22})\,p, & E &= \frac{1}{2}(\alpha_{11} + 2\alpha_{22})\,h, \qquad (1.32) \\
F &= (3\alpha_{22})\,m.
\end{aligned}$$

The equations in (1.19) together with Equation (1.32), after performing integration, yield the following forms of the flow variables:

$$\rho = e^{\frac{\alpha_{11}t}{b}}\,\widehat{S}(\xi), \qquad\qquad v = e^{\delta t}\,\widehat{U}(\xi),$$

$$p = e^{\left(2\delta + \frac{\alpha_{11}}{b}\right)t}\,\widehat{P}(\xi), \qquad h = e^{\frac{1}{2}\left(2\delta + \frac{\alpha_{11}}{b}\right)t}\,\widehat{E}(\xi), \qquad (1.33)$$

$$m = e^{\left(3\delta + \frac{\alpha_{11}}{b}\right)t}\,\widehat{F}(\xi),$$

where

$$\delta = \frac{\alpha_{22}}{b}. \qquad (1.34)$$

Similarity variable ξ, the shock path $\bar{R}(t)$ and the shock velocity are given as

$$\xi = e^{-\delta t}\,r, \qquad \bar{R}(t) = e^{\delta t}, \qquad V = \delta\,e^{\delta t}. \qquad (1.35)$$

Therefore, the flow variables ρ, v, p, h, m, $\rho_0(r)$ and $m_0(r)$ are obtained in the following forms:

$$\rho = \rho_0(\bar{R}(t))S^*(\xi), \qquad\qquad v = V\,U^*(\xi),$$

$$p = \rho_0(\bar{R}(t))V^2\,P^*(\xi), \qquad h = \rho_0^{1/2}V\,E^*(\xi),$$

$$m = m_0(\bar{R}(t))\,F^*(\xi),$$

$$\rho_0(r) = \rho_c\,r^\theta, \qquad\qquad m_0(r) = m_c\,r^\eta, \qquad (1.36)$$

$$\theta = \frac{\alpha_{11}}{\alpha_{22}}, \qquad\qquad \eta = 3,$$

where $S^*(\xi) = \frac{\widehat{S}(\xi)}{\rho_c}$, $U^*(\xi) = \frac{\widehat{U}(\xi)}{\delta}$, $P^*(\xi) = \frac{\widehat{P}(\xi)}{\delta^2\,\rho_c}$, $E^*(\xi) = \frac{\widehat{E}(\xi)}{\delta\rho_c^{1/2}}$ and $F^*(\xi) = \frac{\widehat{F}(\xi)}{m_c}$.

From Equation (1.35), we see that the shock path is exponentially varying. Substituting Equation (1.36) in the system (1.1), we obtain the following system of ODEs after suppressing the asterisk signs

$$S\theta + (U - \xi)S' + SU' + \frac{2U}{\xi}S = 0,$$

$$(U - \xi)U'S + US + P' + \mu EE' + \frac{\mu E^2}{\xi} + \frac{G\,m_c}{\xi^2\,\delta^2}FS = 0,$$

$$(U - \xi)E' + \left(\frac{U}{\xi} + U'\right)E + \left(1 + \frac{\theta}{2}\right)E = 0, \qquad (1.37)$$

$$F' - \frac{4\pi \rho_c \xi^2}{m_c} S = 0,$$

$$(U - \xi)P' + [2 + (1 - \gamma)\theta] P - \gamma P(U - \xi)\frac{S'}{S} = 0,$$

For the strong shock, the R-H conditions are given by

$$U(1) = (1 - \beta), \qquad\qquad\qquad S(1) = \frac{1}{\beta},$$

$$P(1) = \left[(1 - \beta) + \frac{1}{2M_A^2}\left(1 - \frac{1}{\beta^2}\right)\right], \qquad E(1) = \frac{1}{M_A \beta}, \qquad (1.38)$$

$$F(1) = 1.$$

The system (1.37) together with the jump conditions (1.38) may be solved for the shock of infinite strength.

1.5 Imploding Shocks

In this section, we consider an imploding strong shock for the case 1 in neighborhood of implosion, when an imploding shock is about to collapse at the centre or axis, we assume the origin of time t to be the instant at which the shock reaches it, thereby $t \leq 0$ in (1.30). Thus, slightly modification has been made in the expression of the similarity variable using the following transformations:

$$\xi = r/(-t)^\delta, \qquad \bar{R} = (-t)^\delta. \qquad (1.39)$$

Therefore, the intervals of the flow variables become $-\infty < t \leq 0$, $\bar{R} \leq r < \infty$ and $1 \leq \xi < \infty$. At any finite radius r, the gas density, pressure, velocity and the speed of sound are bounded. At the instant of collapse, where $t = 0$ and $\xi = \infty$ for finite r, for the variables u, ρ, p, h and m to be bounded, we have the boundary conditions at $\xi = \infty$ given as follows:

$$U(\infty) = 0, \qquad \frac{P(\infty)}{S(\infty)} = 0, \qquad \frac{E(\infty)}{S(\infty)} = 0. \qquad (1.40)$$

The system (1.30) can be written in the form of matrix as:

$$AW' = B, \qquad (1.41)$$

where $W = (S, U, P, E, F)^{tr}$, and the matrix W, column vector B can be determined from the system (1.30). It may be noticed that the system (1.30) have an unknown parameter δ, which cannot be obtained from the dimensional considerations; it has to be computed from a non-linear eigenvalue problem for a system of ODEs only. Now, the system (1.41) is solved for the variables U', S', P', E' and F' using the Cramer's rule in the following manner:

$$U' = \frac{\triangle_1}{\triangle}, \quad S' = \frac{\triangle_2}{\triangle}, \quad P' = \frac{\triangle_3}{\triangle}, \quad E' = \frac{\triangle_4}{\triangle}, \quad F' = \frac{\triangle_5}{\triangle}, \quad (1.42)$$

where $\triangle$ represents the determinant of matrix A and is given by

$$\triangle = -(U - \xi)^2 \left[(U - \xi)^2 - \frac{(\gamma P + \mu E^2)}{S} \right], \quad (1.43)$$

and $\triangle_k$ ($k = 1, 2, 3, 4, 5$) are the determinants determined by replacing the corresponding kth column of A with column vector B. In the interval $[1, \infty)$, it may be noticed that $U < \xi$. So, $\triangle > 0$ at $\xi = 1$ and $\triangle < 0$ at $\xi = \infty$, which implies that there exists a $\xi \in [1, \infty)$ at which $\triangle$ becomes zero. Therefore, the solutions of the system (1.42) becomes singular. Hence, for obtaining a non-singular solution of (1.41) in the interval $[1, \infty)$, we choose the exponent δ in such a manner that $\triangle$ vanishes at a point, where the determinants $\triangle_1$, $\triangle_2$, $\triangle_3$, $\triangle_4$ and $\triangle_5$ also vanish simultaneously. To obtain such value of exponent δ, we first introduce the variable Z as

$$Z(\xi) = \left[(U - \xi)^2 - \frac{(\gamma P + \mu E^2)}{S} \right], \quad (1.44)$$

whose first derivative, in view of (1.42), is

$$\frac{dZ}{d\xi} = \left(2(U - \xi)(\triangle_1 - \triangle) - \frac{(\gamma \triangle_3 + 2\mu E \triangle_4)}{S} + \frac{(\gamma P + \mu E^2)}{S^2} \triangle_2 \right) / \triangle. \quad (1.45)$$

Therefore, the Equation (1.42), with the help of Equation (1.45), become

$$\frac{dU}{dZ} = \frac{\triangle_1}{\triangle_6}, \quad \frac{dS}{dZ} = \frac{\triangle_2}{\triangle_6}, \quad \frac{dP}{dZ} = \frac{\triangle_3}{\triangle_6}, \quad \frac{dE}{dZ} = \frac{\triangle_4}{\triangle_6}, \quad \frac{dF}{dZ} = \frac{\triangle_5}{\triangle_6}, \quad (1.46)$$

where

$$\triangle_6 = \left(2(U - \xi)(\triangle_1 - \triangle) - \frac{(\gamma \triangle_3 + 2E \triangle_4)}{S} + \frac{(\gamma P + \mu E^2)}{S^2} \triangle_2 \right),$$

with

$$\xi = U + \left(Z + \frac{\gamma P}{S} + \frac{\mu E^2}{S} \right)^{1/2}.$$

1.6 Results and Discussion

On integrating Equation (1.46) for $1 \leq \xi < \infty$ by assigning some value to δ, we obtain the values of S, U, P, E, F and $\triangle_1$ at $Z = 0$. By successive approximations, the value of δ is improved in such a way that for its final value, the determinant $\triangle_1$ vanishes at $Z = 0$. The computed values of δ are listed in Table 1.1 for different values of similarity exponent θ and γ. θ and γ are also satisfy the relation (1.28). It is observed from Table 1.1 that the value of similarity exponent δ decreases with an increase in the value of θ. The values of ρ_c, m_c, M_A and γ have been taken as $\rho_c = 1$, $m_c = 1$, $M_A^{-2} = 0.01, 0.02$, $\gamma = 4/3, 7/5$. The effect of magnetic field on the flow behind the shock are significant when $M_A^{-2} \geq 0.01$ (see [9]).

The expressions of the flow variables, obtained from Equation (1.42), where the shock wave collapse at $t = 0$, $X = 0$ ($t = 0, r \neq 0, \xi = \infty$), are given as

$$S \sim C_1, \quad U \sim \xi^{(\delta-1)/\delta}, \quad P \sim \xi^{2(\delta-1)/\delta}, \quad E \sim \xi^{2(\delta-1)/\delta}, \quad \text{as} \quad \xi \to \infty, \tag{1.47}$$

where C_1 is a constant.

From the expressions in (1.47), we observe that the velocity U and the pressure P tend to zero as $\xi \to \infty$, at the instant collapse due to value of δ smaller than unity, while the density S remains bounded thereat. The system (1.42) is integrated in the range $1 \leq \xi \leq \infty$ and the profiles of the variables ρ, u, p, h and m are illustrated in Figures 1–10. The numerical solutions of flow variables in the neighborhood of $\xi = \infty$ are consistent with the asymptotic results. For numerical calculations, we have taken the values of constant parameters as $\mu = 1.25663753 \times 10^{-6}$ and $G = 6.67408 \times 10^{-11}$.

Table 1.1 Values of Similarity exponent δ for different values of γ and ambient density exponent θ

M_A^{-2}	γ	β	θ	Computed δ
0.01	4/3	0.1658040	0.571250	0.7778317
			1.111110	0.64285737
0.02	4/3	0.1851492	0.593750	0.771084337
			1.100001	0.64516108
0.01	7/5	0.1859448	0.547890	0.78496324
			1.000680	0.666515589
0.02	7/5	0.2104281	0.568750	0.7785888
			1.000050	0.6666555

When we go towards the centre of collapse, from the flow profiles shown in Figures 1–10, we observe that the flow variables density, pressure, magnetic field and the mass increase with an increase in the value of density exponent θ behind the shock whereas the velocity decreases. The gas particle, which passes through the shock leads to the compression of shock, which increases the density, pressure and magnetic field behind the shock and it may also attributed to the area contraction or the convergence of the shock wave. From Table 1.2, we observe that the value of β (density ratio) across the shock front increases with an increase in the value of γ i.e. the shock strength decreases. All the flow variables increase γ increases (see Figures 1.1–1.5).

Table 1.2   Values of density ratio β across the shock front for different values of γ and M_A^{-2}

γ	M_A^{-2}	β
	0	0.1428571
4/3	0.01	0.1658040
	0.02	0.1851492
	0	0.1666667
7/5	0.01	0.1859448
	0.02	0.2104281
	0	0.2481203
5/3	0.01	0.2592906
	0.02	0.2700677

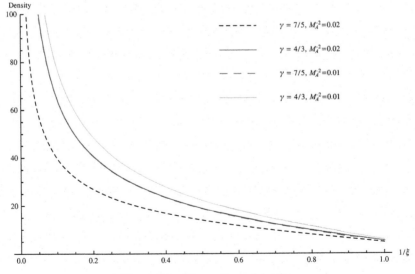

Figure 1.1   Flow pattern of density.

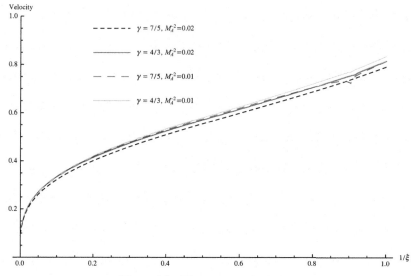

Figure 1.2 Flow pattern of velocity.

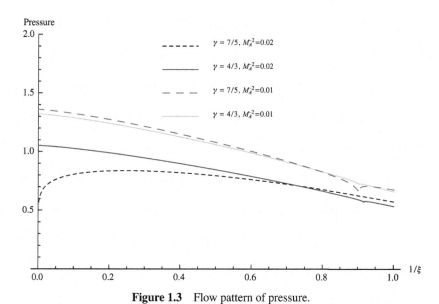

Figure 1.3 Flow pattern of pressure.

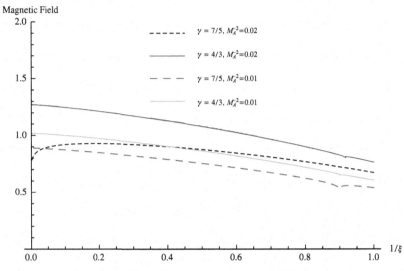

Figure 1.4   Flow pattern of magnetic field.

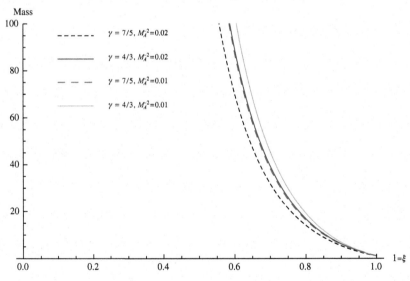

Figure 1.5   Flow pattern of mass.

1.7 Conclusion

In this work, we have used the method of Lie group on the system of equations describing the flow in a self-gravitating gas. By using the invariance condition of Lie group method, we have obtained the determining equations for the system 1.1 and with the help of these determining equations, we have obtained the infinitesimal generators. An ordinary differential equations's system is obtained with the help of infinitesimal generators and then, the system of differential equations together with the jump conditions (1.31) is solved by using the fourth order Runga-Kutta scheme for different values of ambient density θ and for specific heats ratio $\gamma = 4/3,\ 7/5$, which are shown in Figures 1.1–1.5. On the basis of this study, we can conclude that the shock wave, which exist in a self gravitating ideal gas, decreases in strength while propagating towards the centre of collapse.

References

[1] R. Arora, *Similarity Solutions and Evolution of Weak Discontinuities in a Vander Waals Gas*, Canadian Applied Mathematics Quarterly, 13 (2005), pp. 297–311.

[2] R. Arora, A. Chauhan, *Lie Symmetry Analysis and Some Exact Solutions of (2 + 1)-dimensional KdV-Burgers Equation*, International Journal of Applied and Computational Mathematics 5(1)(2019), pp. 1–13.

[3] R. Arora, S. Yadav and M. J. Siddiqui, *Similarity Method for the Study of Strong Shock Waves in Magnetogasdynamics*, Boundary Value Problem, 142 (2014), pp. 1–15.

[4] G. W. Bluman and S. Kumei, *Symmetries and Differential Equations*, Springer Berlin, (1989).

[5] M. Chadha and J. Jena, *Self-Similar Solutions and Converging Shocks in a Non-Ideal Gas With Dust Particles*, International Journal of Non-Linear Mechanics, 65 (2014), pp. 164–172.

[6] G. Deb Ray, *An Exact Analytic Solution for Strong Plane Shock Waves in an Exponential Medium*, Bulletin of Calcutta Mathematical Society, 66 (1974), pp. 27–31.

[7] G. Guderley, *Starke Kugelige Und Zylindrische Verdichtungsstosse in Der Nahe der Kugelmittelpunktes bzw der Zylinderachse*, Luftfahrtforschung 19 (1942), pp. 302–312.

[8] P. E. Hydon, *Symmetry Method for Differential Equations*, A Beginner's Guide, Cambridge University Press, 2000.

[9] P. Rosenau, S. Frankenthal, *Equatorial Propagation of Axisymmetric Megnetohy-Drodynamic Shocks*, Physics of Fluids 19 (1976), pp. 1889–1899.

[10] J. Jena, *Lie Group Transformations for Self-Similar Shocks in a Gas With Dust Particles*, Mathematical Methods in Applied Sciences, 32 (2009), pp. 2035–2049.

[11] J. Jena, *Self-Similar Solutions in a Plasma with Axial Magnetic Field (θ-Pinch)*, Meccanica, 47 (2012), pp. 1209–1215.

[12] J. D. Logan, *Applied Mathematics: A Contemporary Approach*, Willey-Interscience, New York (1987).

[13] G. Nath, J. P. Vishwakarma, V. K. Srivastava and A. K. Sinha, *Propagation of Magnetogasdynamic Shock Waves in a Self-Gravitating Gas With Exponentially Varying Density*, Journal of Theoretical and Applied Physics, 7:15 (2013), pp. 1–8.

[14] G. Nath and S. Singh, *Flow Behind Magnetogasdynamic Exponential Shock Wave in Self-Gravitating Gas*, International Journal of Non-Linear Mechanics, 88 (2017), pp. 102–108.

[15] P. J. Olver, *Applications of Lie Group to Differential Equations*, Springer, New York, (1986).

[16] M. Saxena and J. Jena, *Self-Similar Solutions and Converging Shocks at the Stellar Surfaces*, Astrophysics Space and Science, 362 (2017), pp. 180 (1–11).

[17] V. D. Sharma and Ch. Radha, *Similarity Solutions for Converging Shocks in a Relaxing Gas*, International Journal of Engineering and Science, 33 (1995), pp. 535–553.

[18] V. D. Sharma and R. Arora, *Similarity Solutions for Strong Shocks in an Ideal Gas*, Studies in Applied Mathematics, 114 (2005), pp. 375–394.

[19] K. K. Singh and J. P. Vishwakarma, *Self-Similar Flow of a Mixture of a Non-Ideal Gas and Small Solid Particles Behind a Shock Wave in Presence of Heat Conduction, Radiation Flux and a Gravitational Field*, Mecannica, 48 (2013), pp. 1–14.

[20] J. P. Vishwakarma and G. Nath, *Cylindrical Shock Wave Generated by a Piston Moving in a Non-Uniform Self-Gravitating Rotational Axisymmetric Gas in the Presence of Conduction and Radiation Heat Flux*, In Petrova VM(ed) Advances in Engineering Research, Nova, Hauppauge, 2 (2011), pp. 537–576.

[21] G. B. Whitham, *On the Propagation of Shock Waves Through Regions of Non-Uniform Area of Flow*, Journal of Flud Mechanics, 4 (1958), pp. 337–360.

[22] Y. B. Zel'dovich and P. Raizer Yu, *Physics of Shock Waves and High Temperature Hydrodynamic Phenomena*, Academic Press New York (1967).

[23] G. Nath, *Shock Wave Driven Out by a Piston in a Mixture of Non-Ideal Gas and Small Solid particles under the Influence of the Gravtitation Field With Monocheomatic Radiation*, Chinese Journal of Physics, 56 (2018), pp. 2741–2752.

2

Dual Solutions for Finite Element Analysis of Unsteady Hydromagnetic Stagnation Point Flow of *Cu* – Water Nanofluid Generated by Stretching Sheet

Santosh Chaudhary[*] and KM Kanika

Department of Mathematics, Malaviya National Institute of Technology, Jaipur-302017, India
E-mail: d11.santosh@yahoo.com; kanikatomar94@gmail.com
*Corresponding Author

The present analysis involves the exploration of an unsteady, two-dimensional boundary layer flow of viscous electrically conducting Cu – water nanofluid nearby a stagnation point towards a stretched surface in the presence of a magnetic field. The influences of Navier's slip condition, viscous dissipation, and Joule heating are also considered. Governing nonlinear partial differential equations are transformed into nonlinear ordinary differential equations by employing convenient similarity transformations. A Galerkin finite element method of the linear weighting function is used to solve the reduced boundary layer equations. Fluid flow and fluid temperature for the impacts of solid volume fraction, unsteadiness parameter, magnetic parameter, stretching parameter, velocity slip parameter, and Brinkmann number are investigated and presented in graphical form. The influences of relevant parameters on wall shear stress and the rate of heat transfer are given through the table and discussed in detail. Furthermore, dual solutions obtained in velocity profile, temperature profile, velocity gradient, and heat transfer rate for negative values of unsteadiness parameter.

23

2.1 Introduction

In unsteady flow, properties and conditions associated with the fluid motion depend on the time. It appears in a variety of natural conditions such as flow due to flapping wings of birds, pulsating flow in arteries and flows in the heart. Unsteady flow conditions play a vital role in some engineering areas of hydrodynamics and aerodynamics. Influences of unsteady emerge either over to fluctuations in the surrounding of fluid or over to body self-induced. Study of unsteady flow of non-Newtonian fluid is analyzed by Erdogan and Imrak [1], while Jat and Chaudhary [2] utilized the impacts of viscous dissipation and Joule heating on unsteady flow along to a stretchable surface. Further, several researchers such as Huang and Ekici [3], Chaudhary et al. [4], Yu and Wang [5], Oglakkaya and Bozkaya [6], and Talnikar and Wang [7] investigated about the problems of unsteady flow.

Analysis of the dynamics of electrically conducting fluids namely clued plasma, salted water, liquid metals, and electrolytes is known as magnetohydrodynamics (MHD). Some applications in broad fields of various research disciplines ranging by geo-physical flow to the industrial applications are like as technology of thermo nuclear reactor, power generation reservoirs and lakes, electronic devices cooling process, crystal growth, solar collector, flow meter and metal casting. A magnetic field may cause currents in an electrically conducting fluid in MHD and so fluid creates a Lorentz force that declines the fluid velocity. This type of situation is called a retarding impact of magnetic field on the fluid motion. Akyildiz and Vajravelu [8] developed the boundary layer flow of a viscous fluid in the presence of a magnetic field. Moreover, Jat and Chaudhary [9], Bandyopadhyay and Layek [10], and Mistrangelo and Buhler [11] reported the studies on MHD flow under the different considerations. Some recent investigations on MHD flow with various geometries are consulted by Eegunjobi and Makinde [12], Chaudhary and Choudhary [13], and Ali et al. [14].

The motion of fluid nearby a stagnation point interprets the stagnation flow, which exists for all solid bodies that can be a move in a fluid. Greater pressure, heat transfer, and the mass deposition rates are encountered by the stagnation domain. Basically, cooling of transpiration, solar central accepter uncovered to wind currents, melt spinning process, metals continuous casting and different achievement in hydrodynamic in engineering procedure are the physical implications of the stagnation point flow. Primarily, steady two-dimensional boundary layer flow near a stagnation point due to the stretching area has been presented by Reza and Gupta [15]. Until, Kumari and Nath [16], Khan et al. [17] and Akbar et al. [18], who have discussed extensively on the field of stagnation point flow. Latterly, Chaudhary and Choudhary [19], and Merkin and Pop [20] addressed the recent analysis in this area.

Nanofluid is a material, which is the mixture of solid nanoparticles with the diameter size 1 to 100 nm and the engineered colloidal fluids. Solid nanoparticles are generally taken with the made by metals, carbon nanotubes, oxides, and carbides, while conventional fluid is considered typically as water, ethanol, oil and ethylene glycol. This type of material is used to increase the convection heat transfer accomplishment and to evolve the thermal conductivity of the ordinary fluid. Nanofluid flow has an enormous phenomenon in some concepts such as fibers spinning, metallic sheets cooling in a cooling bath, production of paper, aerodynamic extrusion of rubber sheets, crystal growing and many other topics of industrial applications and technologies. The term nanofluid is created by Choi [21]. After that, Keblinski et al. [22] presented the discussion on the heat flow by the suspension of solid nanoparticles in the fluid. Consequently, nanofluid flow can be quoted by the explorations of Chein and Chuang [23], Turkyilmazoglu [24], Safikhani and Abbasi [25], Sheikholeslami and Rokni [26], and Bezaatpour and Goharkhah [27].

A lot of researchers have created an interest to analyze the fluid flow towards a stretching sheet. This type of flow is useful in many industrial processes, specifically, drawing of wire, paper production, hot rolling, glass fiber, polymer plates manufacturing and polymer extrusion of plastic sheets. Also, stretching flow investigations have done a good effort to boost the information in this area along to the different situations including the assumption of the permeable surface, heat and mass transfer, slip impacts and the MHD flow. The study of thermo-fluid includes significant heat transfer between the surface and the enclosing fluid for the glass fiber and plastic plates manufacturing. There are two ways to improve the mechanical properties of fiber and plastic sheets such as the rate of cooling and the extensibility of the sheet. Accordingly, fluid flow past a nonlinear stretching surface is examined by Vajravelu and Cannon [28]. By varying the stretchable sheet, Jat and Chaudhary [29], Narayana and Sibanda [30], Mustafa et al. [31], and Babu and Sandeep [32] analyzed the flow situations with several configurations. However, very recently, Chaudhary and Choudhary [33], and Aly [34] explored the realistic models by applying the stretching surface effect

The chief goal of this analysis is to extend the study of Malvandi et al. [35] with the considerations of viscous dissipation and Joule heating in the presence of transverse magnetic field along to the Cu − water nanofluid. Galerkin finite element method is utilized to solve the ordinary differential equations, which is obtained by applying the convenient similarity variables.

2.2 Formulation of the Problem

The unsteady two-dimensional flow of copper-water (Cu − water) nanofluid over a stagnation region along a stretchable plate is considered. Nanofluid is taken viscous, incompressible and electrically conducting. There is no thermal equilibrium and no slippage occurs between the base fluid water and the suspended nanoparticles Cu. Values of thermophysical characteristics of used materials are given via Table 2.1 followed by Kalteh [36]. Moreover, a Cartesian coordinate system (x, y) is assumed, where x and y are the coordinates measured along to the surface and perpendicular to it, respectively and fluid is confined in the upper half plane $y \geq 0$ as shown in Figure 2.1. The free stream velocity and the velocity of the stretching plate are $U_\infty = \frac{ax}{(1-ct)}$ and $U_w = \frac{bx}{(1-ct)}$ respectively, where a, b and c are positive constants, and t is the time. It is also surmised that the temperature at the surface and ambient fluid temperature are T_w and T_∞, respectively. The flow is related to a uniform magnetic field with strength H_0, which is applied in the normal direction of the flow. Values of the magnetic Reynolds number is taken small, which tends to neglect the induced magnetic field. Under the above considerations, the governing equations of the continuity, momentum, and energy are

Table 2.1   Thermophysical properties of used materials in present study

Materials	$\kappa\,(W/mK)$	$\rho\,(Kg/m^3)$	$C_p\,(J/KgK)$	$\sigma_e(S/m)$
Water	0.613	997.1	4179	0.05
Cu	400	8933	385	5.96×10

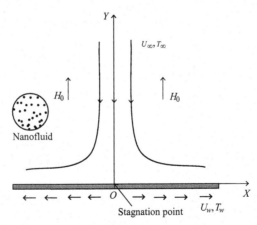

Figure 2.1   Flow configuration.

$$\frac{\partial u}{\partial x} + \frac{\partial v}{\partial y} = 0 \tag{2.1}$$

$$\frac{\partial u}{\partial t} + u\frac{\partial u}{\partial x} + v\frac{\partial u}{\partial y} = \frac{\partial U_\infty}{\partial t} + U_\infty\frac{dU_\infty}{dx} + v_{nf}\left(\frac{\partial^2 u}{\partial x^2} + \frac{\partial^2 u}{\partial y^2}\right)$$
$$-\frac{(\sigma_e)_{nf}\,\mu_e^2 H_0^2}{\rho_{nf}}(u - U_\infty) \tag{2.2}$$

$$(\rho C_p)_{nf}\left(\frac{\partial T}{\partial t} + u\frac{\partial T}{\partial x} + v\frac{\partial T}{\partial y}\right) = \kappa_{nf}\left(\frac{\partial^2 T}{\partial x^2} + \frac{\partial^2 T}{\partial y^2}\right) + \mu_{nf}\left(\frac{\partial u}{\partial y}\right)^2$$
$$+ (\sigma_e)_{nf}\,\mu_e^2 H_0^2\,(u - U_\infty)^2 \tag{2.3}$$

along to the appropriate boundary conditions

$$v = 0, \ u = U_w + v\,N\sqrt{t}\frac{\partial u}{\partial y}, \ T = T_w \text{ at } y = 0$$
$$u \to U_\infty, \ T \to T_\infty \quad \text{as } y \to \infty \tag{2.4}$$

where subscript nf denotes the thermophysical properties of nanofluid, u and v are the velocity components in the x and y directions, respectively. $v = \frac{\mu}{\rho}$ is the kinematic viscosity, μ is the coefficient of viscosity, ρ is the density, σ_e is the electrical conductivity, μ_e is the magnetic permeability, C_p is the specific heat at constant pressure, T is the temperature of nanofluid, κ is the thermal conductivity and N is the initial value of velocity slip factor.

Further, thermophysical characteristics of nanofluid secured to spherical shape solid nanoparticles are coefficient of viscosity, density, electrical conductivity, thermal conductivity, and heat capacitance given by Kalteh [36] are detailed as follows

$$\mu_{nf} = \frac{\mu_f}{(1 - \phi)^{5/2}} \tag{2.5}$$

$$\rho_{nf} = (1 - \phi)\,\rho_f + \phi\rho_s \tag{2.6}$$

$$(\sigma_e)_{nf} = \left\{1 - \frac{3\phi\left[(\sigma_e)_f - (\sigma_e)_s\right]}{2(\sigma_e)_f + (\sigma_e)_s + \phi\left[(\sigma_e)_f - (\sigma_e)_s\right]}\right\}(\sigma_e)_f \tag{2.7}$$

$$\kappa_{nf} = \left[1 - \frac{3\phi\,(\kappa_f - \kappa_s)}{2\kappa_f + \kappa_s + \phi\,(\kappa_f - \kappa_s)}\right]\kappa_f \tag{2.8}$$

$$(\rho C_p)_{nf} = (1 - \phi)\,(\rho C_p)_f + \phi\,(\rho C_p)_s \tag{2.9}$$

where subscripts f and s indicate the physical characteristics of the base fluid and solid nanoparticles, respectively and ϕ is the solid volume fraction of nanoparticles.

2.3 Similarity Transformation

Defining the stream function $\psi(x, y, t)$ such that $u = \frac{\partial \psi}{\partial y}$ and $v = -\frac{\partial \psi}{\partial x}$, which satisfies the continuity Equation (2.1), the similarity variable η and the non-dimensional temperature $\theta(\eta)$ in the following form as (Malvandi et al. [35])

$$\psi = \sqrt{\frac{a\, v_f}{1 - ct}}\, x\, f(\eta) \tag{2.10}$$

$$\eta = \sqrt{\frac{a}{v_f (1 - ct)}}\, y \tag{2.11}$$

$$\theta(\eta) = \frac{T - T_\infty}{T_w - T_\infty} \tag{2.12}$$

where $f(\eta)$ is the dimensionless stream function.

After introducing Equations (2.10) to (2.12) into the momentum and energy Equations (2.2) and (2.3) with the boundary conditions Equation (2.4), reduce to the following non-dimensional nonlinear differential equations

$$\frac{1}{E_1 E_2} f''' + \left(f - \frac{A}{2}\eta\right) f'' - (f' + A)\, f' - \frac{1}{E_2} \frac{(\sigma_e)_{nf}}{(\sigma_e)_f} M\, (f' - 1) + A + 1 = 0 \tag{2.13}$$

$$\frac{\kappa_{nf}}{\kappa_f} \theta'' + E_3 \Pr \left(f - \frac{A}{2}\eta\right) \theta' + Br \left[\frac{1}{E_1} f''^2 + \frac{(\sigma_e)_{nf}}{(\sigma_e)_f} M\, (f' - 1)^2\right] = 0 \tag{2.14}$$

subject to the corresponding boundary conditions

$$\begin{aligned} f = 0, \ f' = \varepsilon + \lambda f'', \ \theta = 1 \ \text{at} \ \eta = 0 \\ f' \to 1, \ \theta \to 0 \quad \text{as} \ \eta \to \infty \end{aligned} \tag{2.15}$$

where $E_1 = (1 - \phi)^{5/2}$, $E_2 = 1 - \phi + \phi\frac{\rho_s}{\rho_f}$, $E_3 = 1 - \phi + \phi\frac{(\rho\, p)_s}{(\rho\, p)_f}$, prime ($'$) denotes the differentiation with respect to η, $A = \frac{c}{a}$ is the unsteadiness parameter, $M = \frac{(\sigma_e)_f \mu_e^2 H_0^2 v_f Re}{\rho_f U_\infty^2}$ is the magnetic parameter,

$Re = \frac{U_\infty x}{v_f}$ is the local Reynolds number, $\text{Pr} = \frac{\mu_f(\ _p)_f}{\kappa_f}$ is the Prandtl number, $Br = \frac{\mu_f U_\infty^2}{\kappa_f(T_w - T_\infty)}$ is the Brinkmann number, $\varepsilon = \frac{b}{a}$ is the stretching parameter and $\lambda = v\,N\sqrt{\frac{at}{v_f(1-ct)}}$ is the velocity slip parameter.

2.4 Local Skin Friction and Local Nusselt Number

The physical quantities of practical interest in this investigation are the local skin friction coefficient C_f and the local Nusselt number Nu, which are given as

$$C_f = \frac{\mu_{nf}\left(\frac{\partial u}{\partial y}\right)_{y=0}}{\frac{\rho_f U_\infty^2}{2}} \tag{2.16}$$

$$Nu = -\frac{x\kappa_{nf}\left(\frac{\partial T}{\partial y}\right)_{y=0}}{\kappa_f(T_w - T_\infty)} \tag{2.17}$$

By using the dimensionless variables Equations (2.10) to (2.12), the physical quantities Equations (2.16) and (2.17) can be defined as

$$C_f = \frac{2}{E_1\sqrt{Re}}f''(0) \tag{2.18}$$

$$Nu = -\frac{\kappa_{nf}}{\kappa_f}\sqrt{Re}\,\theta'(0) \tag{2.19}$$

2.5 Method of Solution

Galerkin finite element method is applied to solve the nonlinear ordinary differential Equations (2.13) and (2.14) with the relevant boundary conditions Equation (2.15). Convenient finite value for the far field boundary conditions $\eta_{\max} = 4$ is assumed for the computational procedure. Introducing

$$f' = h \tag{2.20}$$

then the Equations (2.13) and (2.14) are reduced into the following forms

$$\frac{1}{E_1 E_2}h'' + \left(f - \frac{A}{2}\eta\right)h' - (h + A)\,h - \frac{1}{E_2}\frac{(\sigma_e)_{nf}}{(\sigma_e)_f}M\,(h - 1) + A + 1 = 0 \tag{2.21}$$

$$\frac{\kappa_{nf}}{\kappa_f}\theta'' + E_3 \Pr \left(f - \frac{A}{2}\eta \right)\theta' + Br \left[\frac{1}{E_1}h'^2 + \frac{(\sigma_e)_{nf}}{(\sigma_e)_f}M(h-1)^2 \right] = 0$$

(2.22)

with the reduced boundary conditions

$$f = 0, \ h = \varepsilon + \lambda h', \ \theta = 1 \ \text{at} \ \eta = 0$$
$$h \to 1, \ \theta \to 0 \ \text{as} \ \eta \to \infty$$

(2.23)

The variational form of Equations (2.20) to (2.22) over an individual line element (η_e, η_{e+1}) is express as follows

$$\int_{\eta_e}^{\eta_{e+1}} w_1 \left(f' - h \right) d\eta = 0$$

(2.24)

$$\int_{\eta_e}^{\eta_{e+1}} w_2 \left[\frac{1}{E_1 E_2}h'' + \left(f - \frac{A}{2}\eta \right)h' - (h + A)h \right.$$
$$\left. - \frac{1}{E_2}\frac{(\sigma_e)_{nf}}{(\sigma_e)_f}M(h-1) + A + 1 \right] d\eta = 0$$

(2.25)

$$\int_{\eta_e}^{\eta_{e+1}} w_3 \left\{ \frac{\kappa_{nf}}{\kappa_f}\theta'' + E_3 \Pr \left(f - \frac{A}{2}\eta \right)\theta' \right.$$
$$\left. + Br \left[\frac{1}{E_1}h'^2 + \frac{(\sigma_e)_{nf}}{(\sigma_e)_f}M(h-1)^2 \right] \right\} d\eta = 0$$

(2.26)

where w_1, w_2 and w_3 are weight functions related to functions f, h and θ respectively. The finite element approximations of the functions f, h and θ are summarized as $f = \sum_{j=1}^2 f_j \varphi_j$, $h = \sum_{j=1}^2 h_j \varphi_j$, $\theta = \sum_{j=1}^2 \theta_j \varphi_j$ with $w_1 = w_2 = w_3 = \varphi_i$, $(i = 1, 2)$.

Where φ_i are the shape functions for a typical linear element (η_e, η_{e+1}) and are taken as

$$\varphi_1^{(e)} = \frac{\eta_{e+1} - \eta}{\eta_{e+1} - \eta_e}, \ \varphi_2^{(e)} = \frac{\eta - \eta_e}{\eta_{e+1} - \eta_e}, \ \eta_e \leq \eta \leq \eta_{e+1}$$

Hence, the finite element model of the Equation (2.24) to (2.26) is given by

$$\begin{bmatrix} [K^{11}] & [K^{12}] & [K^{13}] \\ [K^{21}] & [K^{22}] & [K^{23}] \\ [K^{31}] & [K^{32}] & [K^{33}] \end{bmatrix} \begin{bmatrix} \{f\} \\ \{h\} \\ \{\theta\} \end{bmatrix} = \begin{bmatrix} \{b^1\} \\ \{b^2\} \\ \{b^3\} \end{bmatrix}$$

(2.27)

where $[K^{mn}]$ and $[b^m]$ $(m = 1, 2$ and $n = 1, 2)$ are defined as

$$K_{ij}^{11} = \int_{\eta_e}^{\eta_{e+1}} \varphi_i \frac{d\varphi_j}{d\eta} \, d\eta,$$

$$K_{ij}^{12} = -\int_{\eta_e}^{\eta_{e+1}} \varphi_i \varphi_j \, d\eta,$$

$$K_{ij}^{13} = 0,$$

$$K_{ij}^{21} = 0,$$

$$K_{ij}^{22} = \int_{\eta_e}^{\eta_{e+1}} \left\{ -\frac{1}{E_1 E_2} \frac{d\varphi_i}{d\eta} \frac{d\varphi_j}{d\eta} + \left(\bar{f} - \frac{A}{2}\eta \right) \varphi_i \frac{d\varphi_j}{d\eta} \right.$$
$$\left. - \left[\bar{h} + \frac{1}{E_2} \frac{(\sigma_e)_{nf}}{(\sigma_e)_f} M + A \right] \varphi_i \varphi_j \right\} \, d\eta,$$

$$K_{ij}^{23} = 0,$$

$$K_{ij}^{31} = 0,$$

$$K_{ij}^{32} = Br \int_{\eta_e}^{\eta_{e+1}} \left[\frac{1}{E_1} \bar{h}' \varphi_i \frac{d\varphi_j}{d\eta} + \frac{(\sigma_e)_{nf}}{(\sigma_e)_f} M \left(\bar{h} - 2 \right) \varphi_i \varphi_j \right] \, d\eta,$$

$$K_{ij}^{33} = \int_{\eta_e}^{\eta_{e+1}} \left[-\frac{\kappa_{nf}}{\kappa_f} \frac{d\varphi_i}{d\eta} \frac{d\varphi_j}{d\eta} + E_3 \Pr \left(\bar{f} - \frac{A}{2}\eta \right) \varphi_i \frac{d\varphi_j}{d\eta} \right] \, d\eta,$$

$$b_i^1 = 0,$$

$$b_i^2 = -\frac{1}{E_1 E_2} \left(\varphi_i \frac{dh}{d\eta} \right)_{\eta_e}^{\eta_{e+1}} - \int_{\eta_e}^{\eta_{e+1}} \left[\frac{1}{E_2} \frac{(\sigma_e)_{nf}}{(\sigma_e)_f} M + A + 1 \right] \varphi_i d\eta$$

and $b_i^3 = -\frac{\kappa_{nf}}{\kappa_f} \left(\varphi_i \frac{d\theta}{d\eta} \right)_{\eta_e}^{\eta_{e+1}} - \int_{\eta_e}^{\eta_{e+1}} \frac{(\sigma_e)_{nf}}{(\sigma_e)_f} MBr \varphi_i d\eta$ with $\bar{f} = \sum_{i=1}^{2} \bar{f}_i \varphi_i$, $\bar{h} = \sum_{i=1}^{2} \bar{h}_i \varphi_i$, $\bar{h}' = \sum_{i=1}^{2} \bar{h}'_i \varphi_i$.

The entire flow domain is divided into the 1000 equal size linear elements and three functions are to be computed at every node. Therefore, after assembly of all element equations, a matrix of order 3003×3003 is obtained. An iterative process must be utilized for the solution of the constructed linear system. After employing the boundary conditions only the system of 2998 equations remains, which is determined by the Newton-Raphson method while maintaining an accuracy of 10^{-7}.

Table 2.2 Comparison for the computational values of $f''(0)$ and $\theta'(0)$ in Cu − water nanofluid with earlier published work with $M = 0.0$, $\varepsilon = 0.0$, $\lambda = 0.0$, $Pr = 6.2$ and $Br = 0.0$

		$f''(0)$		$-\theta'(0)$	
ϕ	A	Malvandi et al. [35]	Present Results	Malvandi et al. [35]	Present Results
0.1	1	1.76039	1.76051	0.46870	0.46933
0.2		1.82528	1.82549	0.46779	0.46830

2.6 Method Validation

Comparison of the numerical values of $f''(0)$ and $\theta'(0)$ are made with the previously published data Malvandi et al. [35] to measure the precision of the proposed method as the Galerkin finite element method. From Table 2.2, it is concluded that the present results are in good agreement with the earlier published results.

2.7 Numerical Results and Discussion

Computational results of the velocity $f'(\eta)$ and the temperature $\theta(\eta)$ distributions for the impacts of the solid volume fraction ϕ, the unsteadiness parameter A, the magnetic parameter M, the stretching parameter ε, the velocity slip parameter λ and the Brinkmann number Br in Cu − water nanofluid are discussed graphically in this section. Further, the effects of the specified parameters on the surface shear stress $f''(0)$ and the rate of heat transfer $\theta'(0)$ are presented in tabular form and explained in detail.

Effects of the solid volume fraction ϕ on the velocity $f'(\eta)$ and the temperature $\theta(\eta)$ distributions are sketched in Figures 2.2 and 2.3, respectively, taking all other controlling parameters constant. These figures showed that for the enlargement in the solid volume fraction ϕ, the fluid flow and the fluid temperature rise. It is true by the physical analysis that solid nanoparticles aggregate the flow resistance, which step-up the fluid velocity. Subsequently, an increment in the solid volume fraction tends to enhance the nanofluid thermal conductivity, which raises the temperature of the fluid.

Figures 2.4–2.7 show the velocity field $f'(\eta)$ and the temperature field $\theta(\eta)$ for various values of the unsteadiness parameter A, respectively, when the value of additional physical parameters are fixed. Figures 2.4 and 2.6 established that the momentum boundary layer as well as the thermal

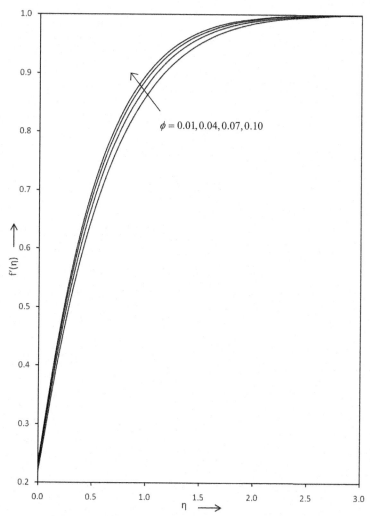

Figure 2.2 Plots of velocity field for various values of ϕ with $A = 0.5$, $M = 0.1$, $\varepsilon = 0.1$ and $\lambda = 0.1$.

boundary layer boost along to the booming values of the unsteadiness parameter A, although the opposite is happening for $\eta > 1.5$ in the momentum boundary layer. Consequently, it is also observed from Figures 2.5 and 2.7 that dual solutions exist for the negative values of the unsteadiness parameter A. From these figures, it is noticed that the fluid flow and the temperature grow-up over to the increment in the unsteadiness parameter A for both

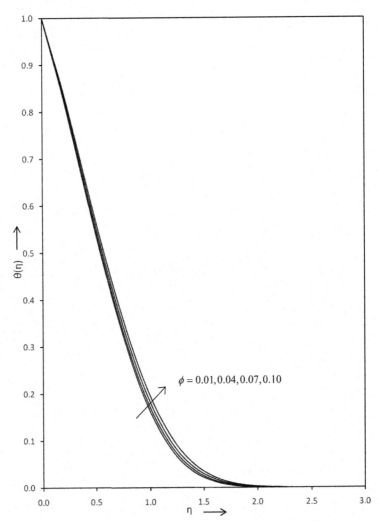

Figure 2.3   Plots of temperature field for various values of ϕ with $A = 0.5$, $M = 0.1$, $\varepsilon = 0.1$, $\lambda = 0.1$, $\mathrm{Pr} = 6.2$ and $Br = 0.62$.

first and second solutions. Until reverse is true for first solution branch in dimensionless velocity if $\eta > 1.4$ and for second solution branch if $\eta > 0.4$, while in the second solution, velocity falls down in installment within the range of η from 0.4 to 0.9. Physically, an enlargement in the unsteadiness parameter implies that surface depreciates the heat, which tends the reduction

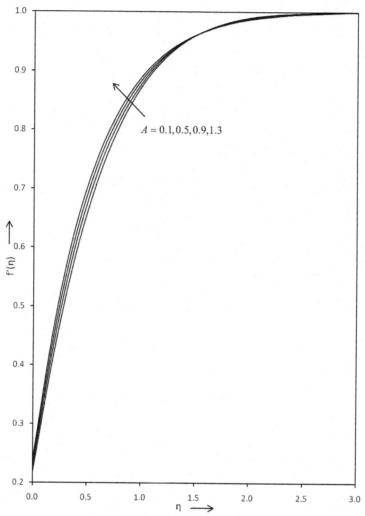

Figure 2.4 Plots of velocity field for various values of A with $\phi = 0.04$, $M = 0.1$, $\varepsilon = 0.1$ and $\lambda = 0.1$.

in temperature, but the thermal boundary layer evolves nearby a stagnation region.

Nature of the dimensionless velocity $f'(\eta)$ and the dimensionless temperature $\theta(\eta)$ along to the impact of the magnetic parameter M are drawn in Figures 2.8 and 2.9 respectively, with the constant value of other parameters. As the value of the magnetic parameter M increases, the velocity of fluid

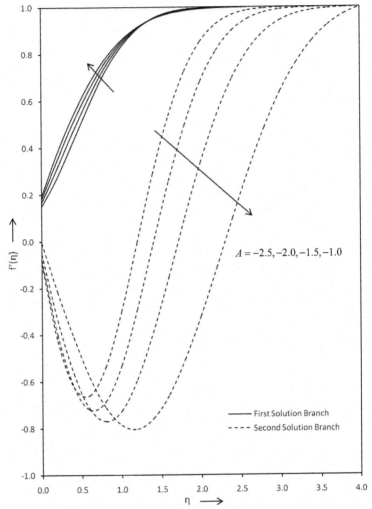

Figure 2.5   Plots of velocity field for various values of A with $\phi = 0.04$, $M = 0.1$, $\varepsilon = 0.1$ and $\lambda = 0.1$.

raises and the temperature of fluid reduces. Because, Lorentz force created for the larger values of the magnetic parameter, which force implies the resistive development into the motion of the fluid and generates more heat resulting in enhancement of the thermal boundary layer.

Figures 2.10 and 2.11 elucidate the influences of the stretching parameter ε on the dimensionless velocity $f'(\eta)$ and the dimensionless temperature $\theta(\eta)$

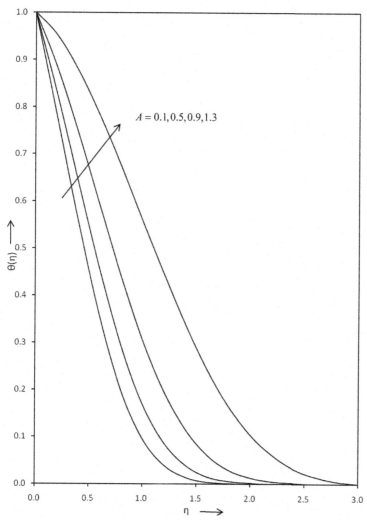

Figure 2.6 Plots of temperature field for various values of A with $\phi = 0.04$, $M = 0.1$, $\varepsilon = 0.1$, $\lambda = 0.1$, $\text{Pr} = 6.2$ and $Br = 0.62$.

respectively while remaining physical parameters are taken as a fixed value. From these figures, it can be noted that the fluid velocity is the increasing function of the stretching parameter ε, although the fluid temperature is the decreasing function of the stretching parameter ε. This happens because the ratio of stretching velocity and the free stream velocity is known as

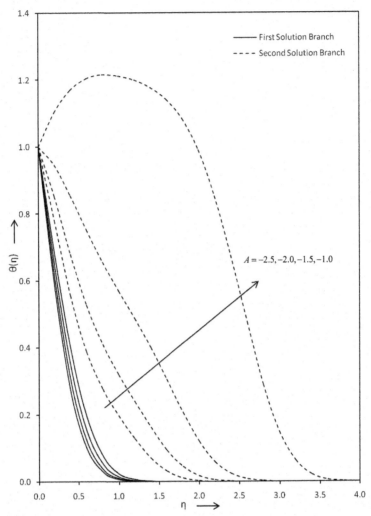

Figure 2.7 Plots of temperature field for various values of A with $\phi = 0.04$, $M = 0.1$, $\varepsilon = 0.1$, $\lambda = 0.1$, $\text{Pr} = 6.2$ and $Br = 0.62$.

the stretching parameter, and the stretching velocity restricts the velocity of free stream and increases rapidly than the velocity of a free stream for raising values of the stretching parameter. Thus, the velocity of fluid evolves and temperature of fluid reduces as the stretching parameter enhances.

Behaviors of changing values of the velocity slip parameter λ on the velocity $f'(\eta)$ and the temperature $\theta(\eta)$ profiles are exhibited via

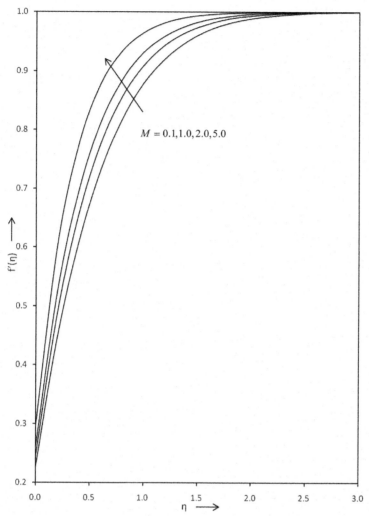

Figure 2.8 Plots of velocity field for various values of M with $\phi = 0.04$, $A = 0.5$, $\varepsilon = 0.1$ and $\lambda = 0.1$.

Figures 2.12 and 2.13 respectively, where the other relevant parameters are constant. It is noted from these figures that an increment in the velocity slip parameter λ implies to develop the velocity, whereas the opposite phenomenon is found in the temperature. This may aspect to the reason that distillation in a stretchable surface is partially connected in the fluid towards

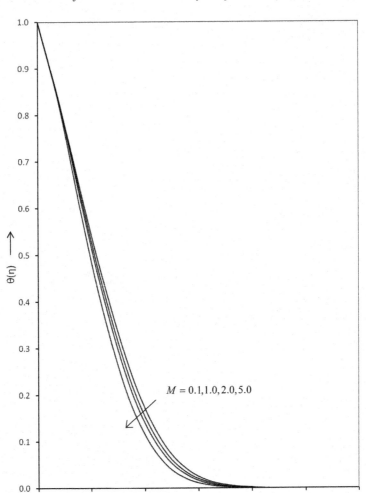

Figure 2.9 Plots of temperature field for various values of M with $\phi = 0.04$, $A = 0.5$, $\varepsilon = 0.1$, $\lambda = 0.1$, $\text{Pr} = 6.2$ and $Br = 0.62$.

the slip condition and heat transfer into the adjacent fluid from the heated surface is in less amount.

Figure 2.14 depicts the influence of the Brinkmann number Br on the temperature $\theta(\eta)$ profile, when the other pertinent parameters kept constant. It is clear from this figure that the temperature rises over to the increasing value of the Brinkmann number Br. This is owing to the impact of the

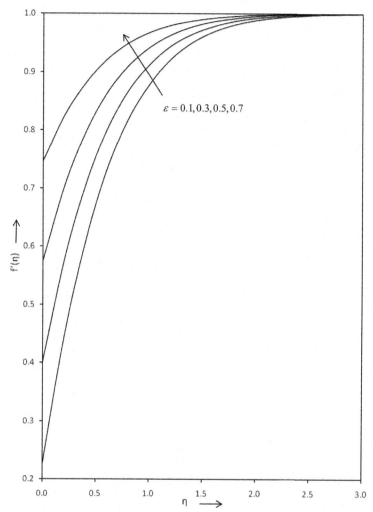

Figure 2.10 Plots of velocity field for various values of ε with $\phi = 0.04$, $A = 0.5$, $M = 0.1$ and $\lambda = 0.1$.

viscous dissipation on the fluid flow leads to the enhancement in energy, that adaptable a larger fluid temperature and also higher buoyancy force. Along to the increasing value of the dissipation parameter, an enlargement in the buoyancy force implies the development in the thermal boundary layer.

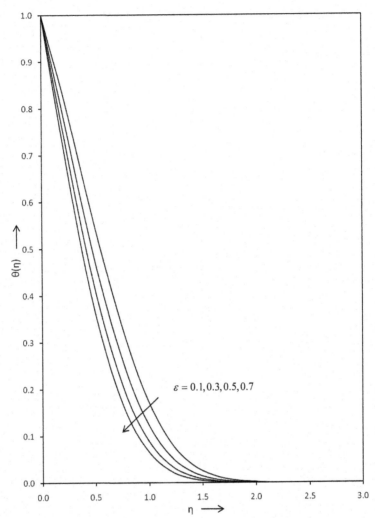

Figure 2.11   Plots of temperature field for various values of ε with $\phi = 0.04$, $A = 0.5$, $M = 0.1$, $\lambda = 0.1$, $\mathrm{Pr} = 6.2$ and $Br = 0.62$.

Table 2.3 exhibits the wall shear stress $f''(0)$ and the rate of heat transfer $\theta'(0)$ for the effects of the solid volume fraction ϕ, the unsteadiness parameter A, the magnetic parameter M, the stretching parameter ε, the velocity slip parameter λ and the Brinkmann number Br in $Cu -$ water nanofluid along to the remaining specified parameters are fixed.

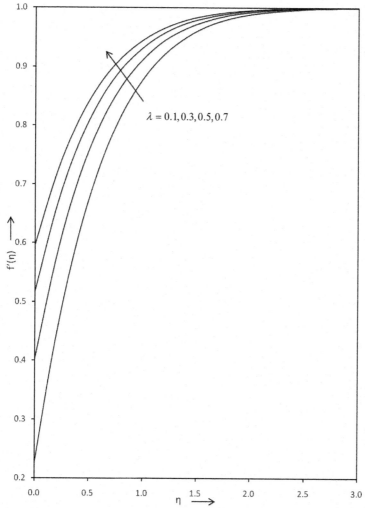

Figure 2.12 Plots of velocity field for various values of λ with $\phi = 0.04$, $A = 0.5$, $M = 0.1$ and $\varepsilon = 0.1$.

It is also observed from Equations (2.18) and (2.19) that $f''(0)$ and $\theta'(0)$ are proportional to the local skin friction C_f and the local Nusselt number Nu respectively. From this table it is noted that along to the enhancing values of the solid volume fraction ϕ, the unsteadiness parameter A and the magnetic parameter M, the wall shear stress and the wall heat

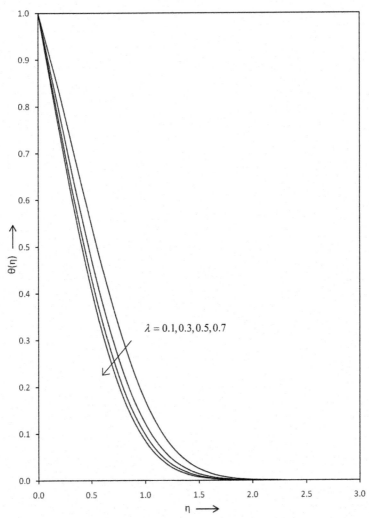

Figure 2.13 Plots of temperature field for various values of λ with $\phi = 0.04$, $A = 0.5$, $M = 0.1$, $\varepsilon = 0.1$, $\mathrm{Pr} = 6.2$ and $Br = 0.62$.

flux develop, while reversal impact is found for the increasingly changing values of the stretching parameter ε and the velocity slip parameter λ. It is also interesting to note that the local Nusselt number rises
for the raise in the Brinkmann number Br. Physically, all negative values of the rate of heat transfer imply that there is a heat flow to the
plate.

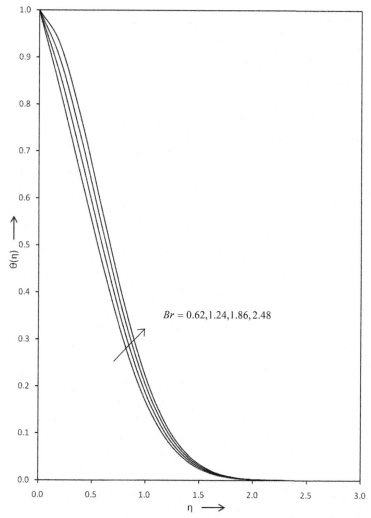

Figure 2.14 Plots of temperature field for various values of Br with $\phi = 0.04$, $A = 0.5$, $M = 0.1$, $\varepsilon = 0.1$, $\lambda = 0.1$ and $Pr = 6.2$.

Dual solutions for the negative values of the unsteadiness parameter A on the wall shear stress $f''(0)$ and the rate of heat transfer $\theta'(0)$ are presented by Table 2.4, keeping other associated parameters fix. According to this table, the surface shear stress and the surface heat flux grow-up along to the increasing value of the unsteadiness parameter A for both solution branches like as first and second solution.

Table 2.3 Numerical data of $f''(0)$ and $\theta'(0)$ for various values of the considering parameters in $Cu -$ water nanofluid with $\mathrm{Pr} = 6.2$

ϕ	A	M	ε	λ	Br	$f''(0)$	$-\theta'(0)$
0.01	0.5	0.1	0.1	0.1	0.62	1.18824	0.78782
0.04						1.25216	0.76331
0.07						1.29938	0.73481
0.10						1.33322	0.70391
0.04	0.1					1.17653	1.01316
	0.9					1.32279	0.47363
	1.3					1.38897	0.13122
	0.5	1.0				1.42605	0.72194
		2.0				1.58636	0.68533
		5.0				1.95172	0.60787
		0.1	0.3			1.00733	1.00807
			0.5			0.74194	1.22201
			0.7			0.45787	1.40427
			0.1	0.3		1.00584	1.00941
				0.5		0.83284	1.15398
				0.7		0.70783	1.24625
				0.1	1.24	1.25219	0.51727
					1.86		0.27117
					2.48		0.02504

Table 2.4 Dual solutions of $f''(0)$ and $\theta'(0)$ for different values of A in $Cu -$ water nanofluid with $\phi = 0.04$, $M = 0.1$, $\varepsilon = 0.1$, $\lambda = 0.1$, $\mathrm{Pr} = 6.2$ and $Br = 0.62$

	$-f''(0)$		$-\theta'(0)$	
	First Solution	Second Solution	First Solution	Second Solution
$-A$	Branch	Branch	Branch	Branch
1.0	−0.93641	1.04512	1.57037	−0.59275
1.5	−0.80759	1.50660	1.78154	0.14006
2.0	−0.66238	1.77090	1.97333	0.57981
2.5	−0.49615	1.91850	2.14785	0.94158

2.8 Conclusions

In this chapter, the influences of the magnetic field and the Navier's slip condition on the boundary layer flow of $Cu -$ water nanofluid nearby a stagnation point over a stretching surface are discussed. Governing boundary layer equations are transformed into coupled high-order non-linear ordinary differential equations by applying the similarity variables. Resulting equations are solved numerically by using the Galerkin finite element method from Matlab software. A summary of the conclusion drawn by the analysis is as follows:

1. Influences of the rising values of the solid volume fraction, the unsteadiness parameter, the magnetic parameter, the stretching parameter and the velocity slip parameter lead to raise the dimensionless velocity, the dimensionless temperature, the local skin friction and the local Nusselt number, whereas reverse effect is found in the fluid flow for $\eta > 1.5$ for the unsteadiness parameter and also reverse phenomenon occur for booming values of the stretching parameter and the velocity slip parameter in the dimensionless temperature, the local skin friction and the local Nusselt number. Further, for the rising values of the magnetic parameter, the temperature of fluid declines.

2. Dual solutions exist for the negative value of the unsteadiness parameter. An enlargement in the unsteadiness parameter implies that the momentum boundary layer, the fluid temperature, the surface shear stress and the surface heat flux are the increasing functions for both solution branches, while momentum boundary layer decreases for $\eta > 1.4$ in first solution branch and also reduces for $\eta > 0.4$ in second solution branch. It is also noted that fluid flow falls partially within the range of η from 0.4 to 0.9 for second solution branch.

3. Thermal boundary layer thickness and the rate of heat transfer evolve with an enhancement in the Brinkmann number.

References

[1] M.E. Erdogan, C.E. Imrak (2007) On some unsteady flows of a non-Newtonian fluid. Appl. Math. Model. 31, 170–180.

[2] R.N. Jat, S. Chaudhary (2009) Unsteady magnetohydrodynamic boundary layer flow over a stretching surface with viscous dissipation and Joule heating. Il Nuovo Cimento 124 B, 53–59.

[3] H. Huang, K. Ekici (2013) An efficient harmonic balance method for unsteady flows in cascades. Aerosp. Sci. Technol. 29, 144–154.

[4] S. Chaudhary, M.K. Choudhary, R. Sharma (2015) Effects of thermal radiation on hydromagnetic flow over an unsteady stretching sheet embedded in a porous medium in the presence of heat source or sink. Meccanica 50, 1977–1987.

[5] M. Yu, L. Wang (2016) A high-order flux reconstruction/correction procedure via reconstruction formulation for unsteady incompressible flow on unstructured moving grids. Comput. Fluids 139, 161–173.

[6] F.S. Oglakkaya, C. Bozkaya (2018) Unsteady MHD mixed convection flow in a lid-driven cavity with a heated wavy wall. Int. J. Mech. Sci. 148, 231–245.

[7] C. Talnikar, Q. Wang (2019) A two-level computational graph method for the adjoint of a finite volume based compressible unsteady flow solver. Parallel Comput. 81, 68–84.

[8] F.T. Akyildiz, K. Vajravelu (2008) Magnetohydrodynamic flow of a viscoelastic fluid. Phys. Lett. A 372, 3380–3384.

[9] R.N. Jat, S. Chaudhary (2010) Radiation effects on the MHD flow near the stagnation point of a stretching sheet. Z. Angew. Math. Phys. 61, 1151–1154.

[10] S. Bandyopadhyay, G.C. Layek (2012) Study of magnetohydrodynamic pulsatile flow in a constricted channel. Commun. Nonlinear Sci. Numer. Simul. 17, 2434–2446.

[11] C. Mistrangelo, L. Buhler (2015) Magnetohydrodynamic flow in ducts with discontinuous electrical insulation. Fusion Eng. Des. 98−99, 1833–1837.

[12] A.S. Eegunjobi, O.D. Makinde (2017) Irreversibility analysis of hydromagnetic flow of couple stress fluid with radiative heat in a channel filled with a porous medium. Results Phys. 7, 459–469.

[13] S. Chaudhary, M.K. Choudhary (2018) Finite element analysis of magnetohydrodynamic flow over flat surface moving in parallel free stream with viscous dissipation and Joule heating. Eng. Comput. 35, 1675–1693.

[14] M.M. Ali, M.A. Alim, S.S. Ahmed (2019) Oriented magnetic field effect on mixed convective flow of nanofluid in a grooved channel with internal rotating cylindrical heat source. Int. J. Mech. Sci. 151, 385–409.

[15] M. Reza, A.S. Gupta (2005) Steady two-dimensional oblique stagnation-point flow towards a stretching surface. Fluid Dyn. Res. 37, 334–340.

[16] M. Kumari, G. Nath (2009) Steady mixed convection stagnation-point flow of upper convected Maxwell fluids with magnetic field. Int. J. Non-Linear Mech. 44, 1048–1055.

[17] Y. Khan, A. Hussain, N. Faraz (2012) Unsteady linear viscoelastic fluid model over a stretching/shrinking sheet in the region of stagnation point flows. Sci. Iran. 19, 1541–1549.

[18] N.S. Akbar, Z.H. Khan, S. Nadeem (2014) The combined effects of slip and convective boundary conditions on stagnation-point flow of CNT suspended nanofluid over a stretching sheet. J. Mol. Liq. 196, 21–25.

[19] S. Chaudhary, M.K. Choudhary (2016) Heat and mass transfer by MHD flow near the stagnation point over a stretching or shrinking sheet in a porous medium. Indian J. Pure Appl. Phys. 54, 209–217.

[20] J.H. Merkin, I. Pop (2018) Stagnation point flow past a stretching/shrinking sheet driven by Arrhenius kinetics. Appl. Math. Comput. 337, 583–590.

[21] S.U.S. Choi (1995) Enhancing thermal conductivity of fluids with nanoparticles. {Publ Fed} 231ASME, 99–106.

[22] P Keblinski, S.R Phillpot, S.U.S Choi, J.A. Eastman (2002) Mechanisms of heat flow in suspensions of nano-sized particles (nanofluids). Int. J. Heat Mass Transf. 45, 855–863.

[23] R. Chein, J. Chuang (2007) Experimental microchannel heat sink performance studies using nanofluids. Int. J. Therm. Sci. 46, 57–66.

[24] M. Turkyilmazoglu (2012) Exact analytical solutions for heat and mass transfer of MHD slip flow in nanofluids. Chem. Eng. Sci. 84, 182–187.

[25] H. Safikhani, F. Abbasi (2015) Numerical study of nanofluid flow in flat tubes fitted with multiple twisted tapes. Adv. Powder Technol. 26, 1609–1617.

[26] M. Sheikholeslami, H.B. Rokni (2017) Influence of melting surface on MHD nanofluid flow by means of two phase model. Chinese J. Phys. 55, 1352–1360.

[27] M. Bezaatpour, M. Goharkhah (2019) Three dimensional simulation of hydrodynamic and heat transfer behavior of magnetite nanofluid flow in circular and rectangular channel heat sinks filled with porous media. Powder Technol. 344, 68–78.

[28] K. Vajravelu, J.R. Cannon (2006) Fluid flow over a nonlinearly stretching sheet. Appl. Math. Comput. 181, 609–618.

[29] R.N. Jat, S. Chaudhary (2008) Magnetohydrodynamic boundary layer flow near the stagnation point of a stretching sheet. Il Nuovo Cimento 123 B, 555–566.

[30] M. Narayana, P. Sibanda (2012) Laminar flow of a nanoliquid film over an unsteady stretching sheet. Int. J. Heat Mass Transf. 55, 7552–7560.

[31] M. Mustafa, J.A. Khan, T. Hayat, A. Alsaedi (2015) Analytical and numerical solutions for axisymmetric flow of nanofluid due to non-linearly stretching sheet. Int. J. Non-Linear Mech. 71, 22–29.

[32] M.J. Babu, N. Sandeep (2016) Three-dimensional MHD slip flow of nanofluids over a slendering stretching sheet with thermophoresis and Brownian motion effects. Adv. Powder Technol. 27, 2039–2050.

[33] S. Chaudhary, M.K. Choudhary (2018) Partial slip and thermal radiation effects on hydromagnetic flow over an exponentially stretching surface with suction or blowing. Therm. Sci. 22, 797–808.

[34] E.H. Aly (2019) Dual exact solutions of graphene-water nanofluid flow over stretching/shrinking sheet with suction/injection and heat source/sink: Critical values and regions with stability. Powder Technol. 342, 528–544.

[35] A. Malvandi, F. Hedayati, D.D. Ganji (2014) Slip effects on unsteady stagnation point flow of a nanofluid over a stretching sheet. Powder Technol. 253, 377–384.

[36] M. Kalteh (2013) Investigating the effect of various nanoparticle and base liquid types on the nanofluids heat and fluid flow in a microchannel. Appl. Math. Model. 37, 8600–8609.

3

Multiparametric Modeling of Carbon Cycle in Temperate Wetlands for Regional Climate Change Analysis Using Satellite Data

**Anna Kozlova, Lesia Elistratova, Yuriy V. Kostyuchenko[*],
Alexandr Apostolov and Igor Artemenko**

Scientific Centre for Aerospace Research of the Earth, National Academy of
Sciences of Ukraine, Kyiv, Ukraine
E-mail: ak.koann@gmail.com; tkach_lesya@ukr.net;
yuriy.v.kostyuchenko@gmail.com; alex@casre.kiev.ua;
igor.artemenko@casre.kiev.ua
*Corresponding Author

This chapter is aimed at regional and local verification of global models of the methane cycle. Models of the atmospheric balance of CH_4 are considered, key emission sources are estimated, and possibilities of forecasting of methane atmospheric concentrations using satellite observations are demonstrated. Partial refinement of the modeling approaches to assessment of emissions components is proposed based on the results of local observations. Models variables, which can be controlled using satellite tools, were determined. Key trends of methane atmospheric concentrations over Ukraine during the observation period have been determined on the base of analysis of SCIAMACHY/Envisat and AIRS & AMSU/Aqua satellites data in the context of observed climate changes. Basic spatial and temporal parameters of the distributions were assessed. Maps of the distributions are prepared and demonstrated. Calculated distributions have been coupled interpreted with data on the distribution of MODIS-generated vegetation and water spectral indices in the framework of constructed models of methane balance. As the result of integrated interpretation, the conclusion on the key drivers of atmospheric methane dynamics over Ukraine are proposed, as well as the conclusions on the improvement of modeling and monitoring methods.

3.1 Introduction

Among the others applied mathematical problems a decision making for the risk analysis is one of most urgent and complicated [1]. A correct and comprehensive solution of this complex problem often requires modeling of impacts and drivers of the disastrous processes [2]. In particular, to analyze a complex ecological risk or climate-related risks we need to model many natural and human-made processes, to estimate corresponding uncertainties and errors [2, 3]. As the many studies demonstrate, correct carbon cycle analysis is a key component of correct climate change scenarios, and so the systems of climate-related risk assessment both on a local and regional scale [3, 4]. And if the cycle of the carbon dioxide we may calculate key parameters with the suitable reliability and accuracy, the methane cycle still essentially uncertain, especially on the local level [5].

Methane (CH_4) is the second most important anthropogenic greenhouse gas, currently contributing about 30% the anthropogenic radiative forcing of CO_2 [6].

Dominant sources of CH_4 to the atmosphere include fossil fuel extraction and use, wetlands, ruminants, rice agriculture, and landfills [6].

According to the National Inventory Report [7], in Ukraine methane emission is about 20% of the total emission. Data of the Report shows that national methane emission is continuously decreasing. But, at the same time, the report declares, that uncertainty of data (aleatoric uncertainty) is about 30–35%, and uncertainty of forming parameters (epistemic uncertainty) is 150%. Such a situation requires the development of approaches and models to reduce uncertainties and to improve the assessments. This is a complex task, includes using novel instruments, utilization of different data sources, and application of sophisticated models.

Different sources of emissions have a different nature and characterize by varied reliability. Industrial sources are the object of traditional inventory, its uncertainty is controlled, and the reliability of its analysis depends on the quality of monitoring [3].

At the same time, about 40% of about 500–600 $TgCH_4$ per year emitted to the atmosphere globally is originated in wetlands, and about 5% is oxidized in unsaturated soils. Correctness and reliability of control of this component are depending not only of monitoring quality but of the model's adequacy.

Over the past several decades, the atmospheric CH_4 growth rate has varied considerably, with changes in fossil fuel emissions [8, 9], atmospheric sinks [6], and fertilizer and irrigation management proposed as explanations (Figure 3.1).

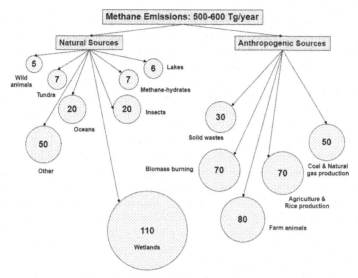

Figure 3.1 Methane estimated sources.

It is important that the methane emissions from terrestrial ecosystems have the potential to form positive feedbacks to climate change.

Wetland ecosystems are of particular concern since they are expected to experience large changes in temperature and precipitation and contain large amounts of potentially labile soil organic matter that is currently preserved by anoxia [10]. The largest wetland complex in the world resides at high latitudes [11] and contributes 10–30% of global CH_4 emissions from natural wetlands [12], but the contribution of moderate wetlands still estimated not enough accurate, especially in view of observed climate changes in the moderate zone.

Interactions of wetland systems with expected 21st century climate change could result in changing CH_4 emissions through several mechanisms:

- thawing permafrost releasing currently dormant soil carbon for degradation [10] and altering surface hydrology via thermokarst [13];
- changes in the kinetics of soil biogeochemistry with increasing temperature [14];
- changes in hydrology interacting with peat properties and active layer depth [15]; and
- changes in net primary productivity (NPP) and plant type distributions [16].

In addition to these decadal-scale changes, the net terrestrial CH_4 surface flux depends non-linearly on very dynamic (i.e., time scale on the order of an hour) interactions between CH_4 production; CH_4 oxidation; aqueous, gaseous, and aerenchyma transport; acid and redox chemistry; and the distribution of soil and surface water. This complex set of interactions and dependence on system properties is difficult to characterize and model globally, making current large-scale CH_4 emission estimates uncertain.

However, the potentially large climate forcing associated with changes in CH_4 emissions motivates the development of land models capable of characterizing these processes and their interactions with the atmosphere.

Taking into account importance of the task, the wide range of the models were developed: from relatively simple regressions of net CH_4 fluxes based on soil properties and climate [16, 17] to the complex models that include details of the various microbial populations that produce and consume CH_4 in the column and their interactions with substrates, pH, and redox potential, in addition to some treatment of the geometry of the rhizosphere and soil horizontal heterogeneity [14].

Also, a wide range of bottom-up estimates of current CH_4 fluxes exists in the literature. For example, recent study estimates average annual emission on the level 31–106 $TgCH_4$ for high-latitudes (north of 50°N), and between 35 and 184 $TgCH_4$ for tropical systems. Also, according to [18] aerobic CH_4 production in living trees could represent a source of 62–236 $TgCH_4$ annually.

High uncertainties of these estimations require harnessing additional data, for example, satellite observation and satellite oriented models. Using the satellite observations of atmospheric CH_4 concentrations [19], allow to estimate that wetlands contribute up to 60% of global emissions and that 2003–2007 CH_4 emissions increase to 7% as a result of global warming.

So, the task of construction of satellite observation oriented model of methane emission from moderate ecosystems, including wetlands, is important. This model should be linked with approaches to anthropogenic emissions inventory, and be oriented to the utilization of satellite observations, as well as include climate change on the local and regional scale.

It allows to clarify regional climate scenarios, rectify adaptation strategies, in particular, in the field of ecological security.

3.2 On the Methodology of Emission Analysis

3.2.1 Generalized Approach

Usually, the task of emission analysis is reduced to the task of assessment of the quantity of carbon emissions [20].

However, the carbon emission coefficients are known with sufficient accuracy only for carbon dioxide emissions. In this case for emission inventory of other GHG such as CH_4, N_2O, NO_X, CO and non-methane hydrocarbons (NMVOC), which connected with corresponding fuel utilization, we use analysis and calculation in separate economical sectors with semi-empirical dependences and statistics on the quantity of burned fuel and emission factors of these gases. Emissions of these gases determined by [20]:

$$E_x = mTK_x \tag{3.1}$$

where E_x – emission of x GHG, m – mass of utilized fuel; T – fuel calorific value, K_x – emission coefficient of the corresponding GHG.

The main types of fuel, which are sources of emissions of these gases are coal, natural gas, oil (gasoline, diesel) and biomass (wood, wood wastes, biofuel). Inventory for emissions can be carried out as well as by the separate types of fuel for all economy sectors, or by the separate sectors for all types of fuel.

Usually, most attention is concentrated on the emissions of carbon dioxide and methane.

Mass of carbon in the utilized fuel m_C we could calculate using the formula:

$$m_C = mTk_C \tag{3.2}$$

where m – mass of utilized fuel; T – calorific value of fuel, which is criteria of fuel thermal capacity; k_C – emission coefficient, which defines actual content of carbon in utilized fuel. It should be noted, that during burning, not all carbon is oxidized. Oxidization rate depends on equipment efficiency, so in mathematical models, we should make a correction to incomplete oxidization of carbon. For it we use the coefficient of carbon oxidization fraction f_C – the relation of the quantity of burned carbon to total carbon content in the fuel:

$$E_{CO_2} = 44m_Cf_C/12 \tag{3.3}$$

Also, we should take into account that significant part of energy resources will not be utilized as fuel. Some part of it will be utilized for the production of other industrial plastic, road bitumen etc. Carbon oxidation, and so, a direct emission in these processes is absent. This carbon calling accumulated and excluded from calculations of total emissions of carbon dioxide. IPCC methodology recommends make a correction to accumulated carbon for the account of some types of fuel – m_{SC} [20]:

$$m_{SC} = f_{SC}m_C \tag{3.4}$$

where f_{SC} – is the fraction of accumulated carbon.

Therefore, a total emission of carbon dioxide reduced to accumulated carbon, we can calculate:

$$E_{CO_2} = 44\,(m_C - m_{SC})\,f_C/12 \tag{3.5}$$

This is the way to calculate anthropogenic emissions of carbon dioxide.

Emissions of CH_4, connected with relevant sources (such as coal mining) traditionally calculating as:

$$E_{CH_4} = 0,67 \cdot mk_{CH_4} \tag{3.6}$$

where m – mass of extracted product (for example, coal in tons), k_{CH_4} – emission coefficient, which defines the volume of gas (m^3) emitted with the extraction of one ton of product, and depends on production technology. To calculation mass of methane (Gg) from its volume usually using the value of the methane density under the pressure of one atmosphere and with the temperature $20°C$, which is equal $0,67 \cdot 10^{-6}$ Gg/m^3.

Total anthropogenic emission, in this case, can be calculated using a simple algorithm:

$$E_{tot} = \sum_{n=1}^{N} \sum_{m=1}^{M} k_{mn} \Delta x_{nm} \tag{3.7}$$

where M – sectors of the economy, N – sites of territory, x – statistical data describing impacts and emissions.

Using algorithm (3.7) we can calculate GHG emissions from every fuel. Besides, data x (in particular, m as the reported mass of extracted fuel) in general case is the function of time, which describe technological and socio-economical development of regions. In this context algorithm (3.7) make a possible to forecast future emissions on the base of future development (see, for example, "emission scenarios" of IPCC).

3.2.2 On the Uncertainty Control

Described approach has significant uncertainties, which depends on the quality of used statistics on the sectoral economy. These uncertainties are aleatoric and connected with the inaccuracy of the data collected and with the correctness of methodology of surveys and data processing.

In this case, it is possible to apply a quite simple method to define errors and uncertainties of the relevant estimations. But we should also remember that the distribution of x is not normal, so we should be careful with "average" assessments.

If we have two different (in ideal case - independent) assessment of emission $E^1_{tot,n}$ i $E^2_{tot,n}$ for certain region n from N and all existing sectors of economy m for studied types of GHG i. To estimate the uncertainty of emissions assessments we can propose a form:

$$U_n = (|E^1_{tot,n} - E^2_{tot,n}|)/(E^1_{tot,n}) \qquad (3.8)$$

And integrated uncertainty for a whole region N, and all types of GHG can be calculated as:

$$U_{tot} = \sqrt{\sum_n U_n^2} \qquad (3.9)$$

This approach is quite primitive and essentially limited. Particularly, it is essentially vulnerable toward the quality of input data. Besides, using methodologically disparate big sets of data may lead to high errors and reliability decreasing, especially in the long time periods. So strategic planning in the field of ecological security, economic development, long-term emission reducing, etc., requires more sophisticated and complex algorithms to estimate uncertainty.

3.3 Modeling of the Carbon Cycle

3.3.1 Key Model Variables and Parameters

Correct analysis of aleatoric uncertainty will not be successful without analysis of epistemic uncertainty. Even estimation of measurement accuracy requires, in the ideal case, understanding of what we are measuring. So we should recognize variables and parameters of processes, which determine the transfer of carbon in the Earth system.

So, a key role in GHG emission uncertainty control may play ecological models, in particular, a correct analysis of the carbon cycle. This issue, as well as some ecological models, is presented in [21]. These models are formalized reflection of our understanding of processes of accumulation and transformation of carbon compounds in ecosystems.

The model includes a number of key elements, which are necessary to calculate correct output data for further analysis.

A mathematical model of carbon balance is usually integrated from regional or local components and links, which reflect the mutual influence of neighboring sites. An integrated mathematical model of the regional carbon cycle for region r may be presented as a matrix equation:

$$\frac{dX_r}{dt} = F_r\left(X_r, V_r, y_r, t\right) \qquad (3.10)$$

where F_r – column matrix of non-linear functions; V_r – column matrix of carbon fluxes between ecosystem components of region r; $X_r = \left\|x_{ra}, X_r^{cl}, X_r^{gl}, X_r^{bl}, X_r^{df}, X_r^{cf}\right\|^T$ – column matrix of unknown variables,

and $X_r^{cl} = \left\|X_r^{clw}, X_r^{cls}, X_r^{clv}, X_r^{clt}, X_r^{clf}\right\|^T$; $X_r^{gl} = \left\|X_{rp}^{glw}, X_{rl}^{glw}, X_{rs}^{glw}\right\|$;
$X_r^{bl} = \left\|X_{rp}^{glw}, X_{rl}^{glw}, X_{rs}^{glw}\right\|$; $X_r^{df} = \left\|X_r^{dfh}, X_r^{dfs}\right\|$; $X_r^{cf} = \left\|X_{rg}^{cf}, X_{rw}^{cf}, X_{rl}^{cf}, X_{rs}^{cf}\right\|$.

Also $X_r^{clw} = \left\|X_{rp}^{clw}, X_{rl}^{clw}, X_{rs}^{clw}\right\|$; $X_r^{cls} = \left\|X_{rp}^{cls}, X_{rl}^{cls}, X_{rs}^{cls}\right\|$;
$X_r^{clv} = \left\|X_{rp}^{clv}, X_{rl}^{clv}, X_{rs}^{clv}\right\|$; $X_r^{clt} = \left\|X_{rp}^{clt}, X_{rl}^{clt}, X_{rs}^{clt}\right\|$; $X_r^{clf} = \left\|X_{rp}^{clf}, X_{rl}^{clf}, X_{rs}^{clf}\right\|$; $X_r^{dfh} = \left\|X_{rg}^{dfh}, X_{rw}^{dfh}, X_{rl}^{dfh}, X_{rs}^{dfh}\right\|$; $X_r^{dfs} = \left\|X_{rg}^{dfs}, X_{rw}^{dfs}, X_{rl}^{dfs}, X_{rs}^{dfs}\right\|$.

Variable y_r, included in this equation, corresponds to carbon fluxes in neighboring regions (with common borders). Value of this variable is proportional to the concentration (content) of carbon in the atmosphere over these neighboring regions as:

$$y_r = \sum_{j=1, j\neq r}^{M} \theta_{rj} X_j^a \qquad (3.11)$$

where θ_{rj} – proportional coefficients; X_j^a – carbon content in the atmosphere over region j.

So, after substituting all expressions, the integrated model of carbon balance for all regions ($r = 1,2,\ldots,M$) may be written as:

$$\frac{dX}{dt} = F\left(X, V, t\right) \qquad (3.12)$$

where $X = \left\|X_1, X_2, \ldots, X_M\right\|^T$ – column matrix of unknown parameters, $V = \left\|V_1, V_2, \ldots, V_M\right\|^T$ – column matrix of carbon fluxes. Number of equations and number of unknown variables in this model, in general case, is equal to the product of ecosystems number n and number of regions M (if all types of ecosystem are presented in every region): $n = 34 \times M = 850$ (for Ukraine quantity of ecosystems type is 34 and number of regions is 25).

Proportional coefficients θ_{rj}, which included into presented equations, determined by links of study region r with neighboring regions j through common borders L_{rj}. These links realized by wind transport of emissions and pollution though region borders. Initial conditions of transportation should be specified by analysis of long-term meteorological measurements: wind rose, frequency and speed of the wind. Sum of vectors of regional wind rose in the study point will indicate as $\vec{w}$. This integrated vector is the function of time, depending on the point of measurement. So, the flux of transported carbon through the border between regions will be calculated as:

$$y_{rj}(t) = X_j^a k_{j\theta} \int_0^{L_{rj}} w(l,t) \cos \psi(l,t) dl \qquad (3.13)$$

where y_{rj} – external toward region r carbon flux from neighboring region j; $k_{j\theta}$ – conversion factor to calculate the average concentration of carbon with height from integrated carbon content in the atmosphere over region j; $\psi(l,t)$ – the angle between the wind rose vector and perpendicular to the r region border. A form of the equation for coefficients θ_{rj} is:

$$\theta_{rj} = k_{j\theta} \int_0^{L_{rj}} w(l,t) \cos \psi(l,t) dl \qquad (3.14)$$

The integrated equation for atmospheric content of carbon may be presented as follows:

$$Y = \Theta X^a \qquad (3.15)$$

where $Y = \|y_1, ..., y_M\|^T$ – column matrix of external carbon fluxes for all regions; $X^a = \|X_1^a, ..., X_M^a\|^T$ – column matrix of carbon contents in the atmosphere over all regions; $\Theta = \|\theta_{rj}\|$, $(r,j = 1,2,...,M)$ – matrix of coefficients. Obviously, $\theta_{rj} \neq 0$ only if regions r and j are neighboring, also is obvious that $\varepsilon \theta_{rj} = -\theta_{jr}$, and matrix Θ is skew-symmetric disperse matrix with zero diagonal. So for task solution, we need only values of elements placed on one side of the diagonal matrix Θ.

Initial Equation (3.1) describing the carbon cycle, is the system with big dimension and big number of variables. From the ecological viewpoint it is possible to divide the studied natural system to sub-systems, in which separate components of studied processes has different velocity. In this case, initial complicated system of the equation may be presented as few more simple sub-systems, which describing separate processes [22]:

$$\frac{dX^a}{dt} = F^a(X,V,Y), \frac{dX^{cl}}{dt} = F^{cl}(X,V,Y),$$

$$\frac{dX^{gl}}{dt} = F^{gl}\left(X, V, Y\right), \frac{dX^{bl}}{dt} = F^{bl}\left(X, V, Y\right), \qquad (3.16)$$

$$\frac{dX^{df}}{dt} = F^{df}\left(X, V, Y\right), \frac{dX^{cf}}{dt} = F^{cf}\left(X, V, Y\right),$$

where F with corresponding indexes is the column matrixes of non-linear functions of fluxes; $X^a = \left\|X_1^a, ..., X_M^a\right\|^T$; $X^{cl} = \left\|X_1^{cl}, ..., X_M^{cl}\right\|^T$. Therefore, we analyze separately equations for carbon fluxes in the atmosphere over all regions, separately for "arable land" ecosystems in all regions, and so on.

A solution of this task can be obtained as the set of discrete values $\tilde{X}_{rv}^a \approx X_r^a\left(t_v\right)$; $\tilde{X}_{rv}^{cl} \approx X_{rv}^{cl}\left(t_v\right)$ on the set of discrete values of time $t_v \in [t_0, t_N]$, where $v = 1,2,...,N$ – the number of integration step; t_0 i t_N – limits of simulation interval. Usually, for modeling, we use an algorithm based on the calculation of carbon content in the atmosphere over some regions. On the every step of modeling new values of unknown variables should be calculated, using an algorithm for atmospheric carbon content calculation:

$$\tilde{X}_{rv+1}^a = X_{rv}^a + h_v \varphi_{rv}^a\left(\tilde{X}_{rv}^a\right), r = 1, 2, ..., M, \qquad (3.17)$$

where h_v – modeling step, φ_{rv}^a – non-linear functions defining by one-step method [22].

Taking into account the low quality of data on atmospheric carbon content obtained from traditional measurements, it important to note, that using in (3.7) regional data from the satellite on the atmospheric concentration of GHG and on corresponding carbon fluxes could increase modeling reliability and simplify integrating procedure.

Other important notice concerns to the necessity of correct problem-oriented land cover classification in study regions. Models show [3] that land cover distribution is very important for correct estimation of carbon balance.

In this carbon balance model for different types of land covers used different variables: X_p^{cli}– carbon in phytomass in ecosystems "arable land" (index i indicates: w – winter crops, s – spring crops, v – vegetable, and t – technical crops), X_p^{gl}– carbon in phytomass in ecosystems "hayfields and pastures"; X_l^{cli} and X_l^{gl} – carbon in litter of these systems; X_s^{cli} Ta X_s^{gl} – carbon in soils of these systems; X_g – carbon in green phytomass (*cf* – conifers, *dfh* – hardwood and *dfs* – softwood), X_w – carbon in wood and root phytomass, X_s – carbon in forest soils. Let V_{ap} – carbon flux from the atmosphere to sub-system "phytomass", which correspond to process of primary production, and V_{ag} – carbon flux from the atmosphere to sub-system "green forest phytomass", which correspond to photosynthesis. It determinates as:

$$V_{ap} = \alpha_{ap} f_{ap} (C, T, P, t) \qquad (3.18)$$

$$V_{ag} = \alpha_{ag} f_{ag} (C, T, P, t) \qquad (3.19)$$

where α_{ap} and α_{ag} – empirical coefficients, determinate by calibration of the model, f_{ap} and f_{ag} – non-linear functions describing dependencies between carbon flux and atmospheric carbon dioxide concentration C, temperature T, precipitation P, and depends on time t, (biophysical processes are mainly cyclical). Thus, implicitly introduced a dependency of photosynthesis intensity from the amount of green biomass.

This model shows that accuracy of classification of land covers and so reliability of technology of interpretation of satellite information aimed to classification is critical for the task of analysis of regional carbon balance. Classification in this task means not only determination of vegetation types, but also analysis of photosynthesis, evapotranspiration, primary production, plant stress, and other parameters. Analysis of land cover distribution, vegetation condition control, and monitoring of atmospheric GHG concentrations using satellite data is the way to increase modeling reliability and correctness though utilization of more correct and comprehensive input data.

3.3.2 From Local to Global Methane Cycle Modeling: Variables and Parameters

If the problem of carbon oxides in the framework of the global and local carbon cycle is considered with sufficient correctness, the problem of methane requiring more detailed analysis. The separate scientific task is to construct a local model describing the methane emission from wetlands, which could be linked with global emission, biospheric and climatic models, such as the corresponding carbon dioxide models (Figure 3.2).

The following way to construct this type of model can be proposed. In the framework of the Community Climate System Model [23] and the Community Earth System Model (CESM1) was constructed CLM4Me model – a CH_4 biogeochemistry model [24].

CLM4 includes modules to simulate plant photosynthesis, respiration, growth, allocation, and tissue mortality; energy, radiation, water, and momentum exchanges with the atmosphere; soil heat, moisture, carbon, and nitrogen dynamics; surface runoff and groundwater interactions; and snow and soil ice dynamics. Having a representation of these processes with some level of

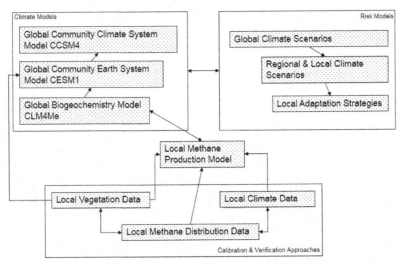

Figure 3.2   Methodology of methane cycle modeling.

detail is important for estimating the controlling factors for CH_4 production, consumption, and emission to the atmosphere. Results of modeling with CLM4Me were used in this study.

For CH_4, the model accounts for production in the anaerobic fraction of soil (P, mol m^{-3}s^{-1}), ebullition (E, mol m^{-3}s^{-1}), aerenchyma transport (A, mol m), aqueous and gaseous diffusion (F_D, mol m^{-2}s^{-1}), and oxidation (O, mol m^{-3}s^{-1}) via a transient reaction-diffusion equation:

$$\frac{\partial(RC)}{\partial t} = \frac{\partial F_D}{\partial z} + P(z,t) - E(z,t) - A(z,t) - O(z,t), \qquad (3.20)$$

here $z(m)$ represents the vertical dimension, $t(s)$ is time, and R accounts for gas in both the aqueous and gaseous phases: $R = \varepsilon_a + K_H \cdot \varepsilon_w$, with ε_a is the air-filled porosity, ε_w is the water-filled porosity, and K_H is the partitioning coefficient for the species of interest.

Because most of the climate models, in particular, CLM4, does not currently have a local wetland representation that includes details relevant to CH_4 production (wetland specific plants, anoxia controls on soil organic matter turnover), nor is the inundated fraction used by the CH_4 submodel integrated with the CLM4 soil hydrology and temperature predictions, this task should be solved using local models and data.

In CLM4Me, CH_4 production in the anaerobic portion of the soil column is related to the grid cell estimate of heterotrophic respiration from soil and litter (R_H; mol Cm^{-2}s^{-1}) corrected for its soil temperature (T_s) dependence, soil temperature through a Q factor (f_T), pH (f_{pH}; [24]), redox potential (f_{pE}), and a factor accounting for the seasonal inundation fraction (S, described below):

$$P = R_H f_{CH_4} f_T f_{pH} f_{pE} S, \qquad (3.21)$$

here, f_{CH_4} is the baseline fraction of anaerobically mineralized C atoms becoming CH$_4$.

Assuming that CH$_4$ production is directly related to heterotrophic respiration implies the assumption that there are no time delays between fermentation and CH$_4$ production, and that soil organic matter can be treated uniformly with respect to its decomposition under either aerobic or anaerobic conditions.

The effect of seasonal inundation is also included in CLM4Me model. It was determined, that in the continuously inundated wetland, anoxia suppresses decomposition, leading to a larger soil organic matter stock, partially compensating the effect of anoxia on decomposition rates. The compensation is complete (respiration rates are unchanged) if i) soil decomposition is a linear function of pool size; ii) the fully anaerobic decomposition rate is a fixed factor of the aerobic rate, and iii) the soil is in equilibrium. At the same time, a tropical seasonally inundated system may experience extensive decomposition during the dry season but emit most of its CH$_4$ during the wet season. The calculation is shown, that because the equilibrium carbon stock will be smaller, the CH$_4$ fluxes will be smaller than the annual wetland even during the wet season.

In CLM4Me, was used a simplified scaling factor to calculate the impact of seasonal inundation on CH$_4$ production:

$$S = (\beta(f - \bar{f}) + \bar{f})/f, S \leq 1, \qquad (3.22)$$

where f is the instantaneous inundated fraction, $\bar{f}$ is the annual average inundated fraction (evaluated for the previous calendar year) weighted by heterotrophic respiration, and β is the anoxia factor that relates the fully anoxic decomposition rate to the fully oxygen-unlimited decomposition rate.

Also, it may be assessed a global CH$_4$ flux and atmospheric methane uptake using inundation phenomena as [25]:

$$\varphi = \frac{1}{1 - \eta \cdot C_{O2}}, \qquad (3.23)$$

here, φ is the factor by which CH_4 production is inhibited above the water level (compared to production as calculated in Equation 31), C_{O2} (mol m^{-3}) is the bulk soil oxygen concentration, and $\eta = 400$ m^3 mol^{-1}.

Using the CLM4Me model it is possible to calculate global distributions of methane emissions in view of climate change. At the same time, the model of wetland is still required.

As the local addition linked to global models, a structural mathematical model of local CH_4 production and oxygenation in wetlands may be proposed.

Model is based on description of acetoclastic methane production [26], where stratification of environment and velocity of substrates input are described as [4]. Model equations can be presented in the combined form [5]:

$$\frac{dx_{ga}}{dt} = x_{ga}(\mu_{ga} - \mu_{dga}), \tag{3.24}$$

$$\frac{dx_{gh}}{dt} = x_{gh}(\mu_{gh} - \mu_{dgh}), \tag{3.25}$$

$$\frac{dx_t}{dt} = x_t(\mu_t - \mu_{dt}). \tag{3.26}$$

Where the velocity of growth of methano-trophic and methano-genetic biomass:

$$\mu_{ga} = \frac{\mu_{\max ga} \cdot s_H}{s_H + K_{sga} + \left(s_H^2/K_i\right)}, \tag{3.27}$$

$$\mu_{gh} = \frac{\mu_{\max gh} \cdot [H^+]}{[H_2] + K_{sgh}}, \tag{3.28}$$

$$\mu_t = \frac{\mu_{\max t} \cdot C_{hl}}{K_{sl} + C_{hl}}, \tag{3.29}$$

$$s_{Ha} = s_a \frac{[H^+]}{K_a + [H^+]}, \tag{3.30}$$

$$\frac{ds_a}{dt} = D_2(s_{in} - s_a) - \frac{\mu_{ga} \cdot x_{ga}}{Y_{xga}} + acc, \tag{3.31}$$

$$\frac{d[H_2]}{dt} = D_2([H_2]_{in} - [H_2]) - \frac{\mu_{gh} \cdot x_{gh}}{Y_{xgh}} + hyd. \tag{3.32}$$

So, the velocity of methane production and consumption:

$$Q_{ch4ga} = kY_{CH4xga}\mu_{ga}x_{ga}, \tag{3.33}$$

$$Q_{ch4gh} = kY_{CH4xgh}\mu_{gh}x_{gh}, \tag{3.34}$$

$$Q_{ch4t} = \mu \frac{x_t}{Y_{xts}}. \tag{3.35}$$

And methane concentration dynamics in upper and lower layers of wetland:

$$\frac{dC_{h1}}{dt} = kg_1(C_{h2} - C_{h1}) - kg_2(C_{h1} - C_a) - Q_{CH4t}, \tag{3.36}$$

$$\frac{dC_{h2}}{dt} = Q_{CH4ga} + Q_{CH4gh} - kg_1(C_{h2} - C_{h1}). \tag{3.37}$$

In these equations:

$x_t, x_{ga}, x_{gh}, \mu_t, \mu_{ga}, \mu_{gh}, s_a, s_{Ha}, [H_2], C_{h1}, C_{h2}, Q_{CH4ga}, Q_{CH4gh}, Q_{CH4t}.$

The solution of these Equations (3.34)–(3.37) allows to fulfill a matrix (3.337) with more correct data on the local scale. The whole approach (3.0)–(3.37) might be used to calculate integrated methane emissions from natural systems in multi-scale range, which is useful for using different measurement tools – in-field gasometry, satellite data, statistics, etc. Using this multi-scale methane wetlands emission model in the tasks of carbon emission analysis is the way to reduce uncertainties, both aleatorical and epistemic.

3.3.3 Uncertainty Analysis in the Carbon Models

For error and uncertainty analysis in integrated models of carbon balance traditionally using data of statistically independent sources. For it may be used data of satellite observations.

In [27] algorithm for determination of standard error $m^2{}_y$ of carbon flux $Y = f(X_i)$, which defined as a function of random distribution X_i with $i = 1, 2, \ldots k$ is proposed in the form:

$$m^2_y = \sum_{i=1}^{k} \left(\frac{dY}{dX_i} m_i \right)^2 + 2 \sum_{i>j} \left(\frac{dY}{dX_i} \right) \left(\frac{dY}{dX_j} \right) r_{ij} m_{X_i} m_{X_j} \tag{3.38}$$

where r_{ij} – coefficients of correlation between distributions X_i and X_j.

Using this algorithm, it is possible to estimate an error of carbon fluxes in all components of the system studied basing on all available data. Satellite observation data is also possible to integrate into this algorithm [3].

Relative integrated reliability of data can be estimated separately. It can be done through comparison of geospatial data of land cover distribution (obtained from satellite image classification), ecological modeling, geostatistics, and inventory, as well as analysis of the atmospheric concentration of

GHG controlled from satellites. For this purpose could be used "agreement index of suitability" [27] as the measure of corresponding of normalized distributions:

$$S_{nm} = \frac{1}{q} \sqrt{\sum_{j=1}^{q} (x_{nj}^{norm} - x_{mj}^{norm})^2}, \text{ where } x_{j}^{norm} = \frac{x_j - x_{j\min}}{x_{j\max} - x_{j\min}} \quad (3.39)$$

where q – descriptive parameters; n – sites of the studied area; m – sets of statistics (for example, departmental data); x – measured values of parameter j.

Mutual consistency of data distribution can be assessed using this method, and so we can more correctly combine disparate data, and minimize errors (aleatoric uncertainty).

Therefore, using of integrated models of carbon balance, especially regionalized with satellite data, allows calculating ecological and climate-related change more accurately. So, modeling is a suitable part of security management.

3.3.4 On the Local Models Integration into the Global Models: A Methodology

The task of refinement of large-scale scenarios based on global models using the local modeling should be considered. This task is fundamentally different from the so-called task of "regionalization" of climatic models. The task of regionalization of models in the traditional sense is usually reduced to the "downscaling" tasks, or to the change (usually – decrease) the spatial and temporal dimension of the distributions obtained from the models. The proposed approach includes refinement of calculations based on the use of additional data sets not included in the standard model (typically, global models include 5–10% of the available measurements in the research area, which is obviously not enough in the calculation of local forecasts).

The principal complexity of this problem lies in the fact that climatic model parameters usually are calculating in the so-called geo-space, or, by definition, in the quasi-two-dimensional space of the earth's surface characterized by closure, positive curvature (convexity), quasi-fractality, and anisotropy in geophysical fields (for example, in the gravitational, magnetic field of the Earth). Geospatial location is determined by spherical (latitude and longitude) or rectangular coordinates. The anisotropy of the geo-space is expressed in the unevenness of the horizontal and vertical directions (vertical movement is considerably complicated), and resulting in sphericity of the Earth's structure, as well as latitudinal and longitudinal directions, the manifestation of which is latitudinal zonation. This determines the specific topology of data allocations for model calculations. In this situation, the results of observations are the sets of data obtained at certain points, and thus, although they are geo-referred, it is not necessarily correspond to all basic postulates that follow from the definition of geo-space.

From a mathematical viewpoint, the problem is to include the observation data (a priori stochastic data, which must be described by completely other methods and which are not necessarily complete) into the set of calculation data with the given topology (into the model data calculated on geo-space). In the general case, solving this task should solve the topological problem of bringing the topology of one set (with a given topology) with the number of the sets of random data with an a priori indefinite topology.

In our case, the topology of the set of model data is given by the definition of geo-space. While the topology of sets of observational data is a priori not defined [28]. Basing on the general theory of obtaining and analyzing data of experiments [29], we can analyze data of current meteorological measurements in terms of the theory of Hilbert spaces, in which limited and unrestricted operators, random elements and operators are defined, and thus the elements of linear and convex programming are determined. Thus, in our case, the use of the theory of finite-dimensional Hilbert spaces for the analysis of meteorological measurements data distributions may allow us to obtain results that will enable us to determine the topology of the corresponding sets and integrate data into sets with the topology, defined in the geo-space.

The representation of observations in Hilbert spaces allows us to stay within the framework of the so-called "local simulation" and thus avoid the need of determination of boundary conditions dependent on the data structure [30]. This frees us from the need to harmonize in the future the boundary conditions of two different types of data. Thus, the mathematical problem is reducing to find the correct method for integrating meteorological observations, in the general case presented in the Hilbert random spaces, and simulation data presented as uniform normalized sets of data defined in the geo-space with the given topology.

Figuratively this task can be represented as the gluing of patches of arbitrary shape and size on the surface of a round ball. In this case, the patches must be adjusted to each other, the total area of the patches is much smaller than the area of the ball, the shape and topology of the patches can be anyone, the ball has a finite elasticity, i.e. it can never change the positive curvature.

The method proposed in this section is that the patches are pre-smoothed, their edges are cut and fitted to each other, after which they are consistently glued to the surface of the ball.

So, we propose to embed into the model the additional data not used during the construction of the model and re-calculate the necessary parameters on a grid that meets the needs of local simulation and is based on available data. This means that, unlike traditional downscaling, the proposed approach is based not on the results of simulation and on the properties of the resulting sets (with a substantially limited spatial and temporal resolution), but on the data of regional measurements. Therefore, the local data becomes decisive for regionalization of the model in this approach, which is a fundamental difference.

Thus, the problem dividing into the problem of reducing the method of observing analysis and obtaining data measurements distributions, and on the task of integrating the obtained reduced data into the model distributions with subsequent calculation of parameters on a reduced grid.

When we solve the problem of specifying the model on the sets of observational data presented as the Hilbert spaces in the part of the reduction of the method of observation analysis, we will consider sets of data $\xi(x_{ij})$, measured in the framework of the method presented by linear operators $A_0 \to A$. To refine A we will be with sufficiently known (with known and/or controlled epistemic reliability) signal $f'(x_{ij})$, represented by a random vector with a known covariance operator $F \in (\Re \to \Re), Fx = Ef(x, f)$, $x \in \Re$, measuring the sets of variables [29]:

$$\xi' = Af' + \nu'. \tag{3.40}$$

Where ν is a random element of a Hilbert space $\tilde{\Re}$ with a correlation operator $\Sigma x = E\nu(x, \nu), x \in \tilde{\Re}, E\nu = 0$, which determines the error, or a "noise" – the measure of aleatoric uncertainty of measurements; ξ' – measurements, refined with the model ξ; E – mathematical expectation; A – methodology or the "model of measurements" are random linear operators (such that, $A \in B \Re \to \tilde{\Re}, x \in \Re, y \in \Re$ while the function $f(Ax, y) = f(x, A * y)$) defining the method of data transformation.

If we introduce any measurable set M ($M \subset \tilde{\Re}_B \Re$), on which we define a random vector η, that describes the distribution of regularized data with a certain reliability, then in relation to it the problem of determining the methodology for integrating observation data into a general model, that is, refinement (reduction) operators, can be defined as [31]:

$$A_\eta = E(A|\eta), J_\eta = (E((A - A_\eta) * F(A - A_\eta) * |\eta). \tag{3.41}$$

Here F is a covariance data operator; J is the operator of uncertainty.

In this case, the task of reducing the data distributions can be formulated as:

$$E(\inf\{E(||R'\xi - Uf||^2|\eta)R' \in (\tilde{\Re} \to U)\}) =$$
$$= E(E(||R_\eta\xi - Uf||^2|\eta)) = E||R_\eta\xi - Uf||^2 \quad , \qquad (3.42)$$

Where U is the orthogonal projector to the subspace $\Re$. Thus, we are talking about the determination of a random operator $R' = R_\eta$, which corresponds to the condition [31]:

$$\inf\{E(||R'\xi - Uf||^2|\eta)R' \in (\tilde{\Re} \to U)\} = E(||R_\eta\xi - Uf||^2|\eta). \quad (3.43)$$

After solving this problem, we obtain mutually coherent distributions of observation data with coordinated boundaries, based on the results of the analysis of data from individual measurements. The resulting sets, represented on Hilbert spaces, can be integrated (reduced) into the global models of any complexity with current topological properties.

Thus, at the first stage, the problem is reduced to the determination of regularized distributions of observation data. In addition, it should be noted that the problem of estimating the uncertainties of the distributions obtained is still urgent.

The task of obtaining the sets of regularized in space and time statistically reliable distributions of required indicators from the observed data (for example, from meteorological stations) in the research area can be solved using the proposed [32] algorithm for nonlinear spatial-temporal data regularization based on the analysis of the main components with the modified method of the smoothing nonlinear kernel-function – Kernel Principal Component Analysis (KPCA) [33]. Using the proposed algorithm we obtain a regularized spatial-temporal distribution of the characteristics of the investigated parameters throughout the observation period with smoothened reliability, taking into account all sources of observation [32, 33].

In accordance with the above general approach, the task is to integrate a plurality of data presented in the form of Hilbert spaces with given linear transformation operators to a set of data with a given topology (a set of model decisions).

So, based on the results of data analysis in the framework of the described method, we obtain a set of normalized distributions $\xi_t = A_t f(x_t) + \nu_t$, where t is the time (defined as the modeling step for a plurality of model data and as a measure of the data set). In the future, it can be proposed to analyze jointly the modeling and observation data in the modified Ensemble Transform Kalman Filter (ETKF) modification method [34].

In this case, we assume that the vectors of the true state of the system x at time k are determined in accordance with the general law:

$$x_t = F_t x_{t-1} + B_t u_t + w_t, \tag{3.44}$$

Where F_t is the matrix of the evolution of the system or the simulated effects on the vector x_{t-1} at the moment $t-1$; B_t is the matrix of controls, measured by the effects u_t on vector x; w_t is a random process with the covariant matrix Q_t. In this way, we introduce the description of the model distributions F_x and the observation data B_t.

Let's determine the extrapolation value of the vector of the true state of the system by evaluating the state vector in the previous step:

$$\hat{x}_{t|t-1} = F_t \hat{x}_{t-1|t-1} + B_t u_{t-1}. \tag{3.45}$$

For this extrapolation value of the vector of the true state one can determine the general form of the covariance matrix:

$$P_{t|t-1} = F_t P_{t-1|t-1} F_t^T + Q_{t-1}. \tag{3.46}$$

The difference between the estimated (extrapolation) value of the vector of the true state of the system and obtained at the appropriate simulation step can be estimated as:

$$\Delta \hat{x}_t = \xi_t - A_t \hat{x}_{t|t-1}, \tag{3.47}$$

A covariance matrix of deviation:

$$S_t = A_t P_{t|t-1} A_t^T + R_t. \tag{3.48}$$

Then, based on the covariance matrices of extrapolation to the state vector and the measurements, we introduce a matrix of optimal coefficients of the Kalman amplification:

$$K_t = P_{t|t-1} A_t^T S_t^{-1}. \tag{3.49}$$

Using this we will adjust the extrapolation values of the vector of the true state of the system:

$$\hat{x}_{t|t} = \hat{x}_{t|t-1} + K_t \Delta \hat{x}_t. \tag{3.50}$$

Also, we introduce a geo-referred filter for the distribution of the vector of the state x_{ij}, which will depend on geographically bound j, i and this will not depend in the general case on the time t.

$$(x_{ij})_t = (x_{ij})_t^\alpha = (x_{ij})_t \alpha_{ij}, \tag{3.51}$$

Here, the coefficients α are chosen according to the introduced KPCA algorithm [35], according to the rule of estimating the optimal balance of the mutual validation function and the covariance matrix:

$$C^F v = \frac{1}{N} \sum_{j=1}^{N} \Phi(x_j)\Phi(x_j)^T \cdot \sum_{i=1}^{N} \alpha_i \Phi(x_i), \qquad (3.52)$$

Where non-linear function of the data distribution Φ is corresponded to conditions $\sum_{k=1}^{N} \Phi(x_k) = 0$, and $\tilde{k}_t$ is the average value of kernel matrix $K \in R^N$ ($[K]_{ij} = [k(x_i, x_j)]$). This matrix consists of kernel vectors $k_t \in R^N$, $[k_i]_j = [k_t(x_t, x_j)]$, and calculating according to modified rule $k_t(x_i, x_t) = \left\langle \rho_{j,t}^{x_j}(1 - \rho_{j,i})^{x_j} \right\rangle$, where ρ – empirical coefficients, selected from the model of studied phenomena [32].

If we apply this filter and remember to reconcile the sets of data, we can offer the form of the covariance matrix P^a to analyze the actual errors based on the form of the covariance matrix of the extrapolated value of the state vector of the P^f system and the matrix of covariance of the observational data R:

$$P^a = P^f - P^f A^T (A P^f A^T + R)^{-1} A P^f. \qquad (3.53)$$

Thus we obtain a tool for optimizing the calculation of the matrix of optimal coefficients of the Kalman amplification and, accordingly, correction of the extrapolation values of the vector of the true state of the system on the aggregate data of modeling and observations.

After completing the data integration procedure, we can calculate the necessary parameters by algorithm [33]:

$$x_t^{(ij)} = \sum_{m=1}^{n} w_{ij}(\hat{x}_t^m) x_t^m, \qquad (3.54)$$

Where $w_{ij}(\tilde{x}_t^m)$ is the weighing coefficient, determining though the rule [33, 35]:

$$\min \left\{ \sum_{m=1}^{n} \sum_{x_t^m \in R^m} w_{ij}(\tilde{x}_t^m) \left(1 - \frac{x_t^m}{\hat{x}_t^m} \right)^2 \right\}. \qquad (3.55)$$

In this equation, m is the number of experiments conducted; n is the number of data sources; x_t^m distribution of the results of observations; R^m – set of data; $\hat{x}_t^m$ is the corrected extrapolation values of the vector of the true state of the system by aggregate modeling and observational data.

Therefore, we obtain a regular spatial distribution of measurable characteristics in the local study area, both as a result of model calculations, and taking into account regional meteorological measurements on a grid that corresponds to the distribution of measurement data, that is, it has a much better resolution than the usual model.

3.4 Satellite Tools and Data for the Carbon Cycle Control

3.4.1 Satellite Tools for GHG Monitoring

As an example of an algorithm for atmosphere composition and conversion radiation parameters to separate gases concentration, we can use approach [36, 37] developed for SCIAMACHY sensor. This algorithm (Weighting Function Modified Differential Optical Absorption Spectroscopy – WFM-DOAS) allows obtaining concentrations of CO_2, CH_4 and O_2 using satellite observations data [37].

This algorithm is based on the least squares method for downscaling of selected vertical atmospheric profiles. Parameters are defining by direct observations of vertical characteristics of the atmosphere. The logarithm of linearized radiation transfer, which is defined by the atmosphere model and observation data, together with polynomial $P_i(a)$ is reducing to the logarithm of the relation of satellite measured radiation and solar radiation (i.e. to the normalized solar radiation I^{obs}). The integrated equation of WFM-DOAS algorithm can be presented as [36]:

$$\sum_{i=1}^{m} \left(\ln I_i^{obs} - \ln I_i^{mod}(\hat{V}, \hat{a}) \right)^2 \equiv \|RES\|^2 \to \min \qquad (3.56)$$

In this equation the model of linear radiation transport is presented as:

$$\ln I_i^{mod}(\hat{V}, \hat{a}) = \ln I_i^{mod}(\bar{V}) + \sum_{j=1}^{J} \frac{\partial \ln I_i^{mod}}{\partial V_j} \bigg|_{\bar{V}_j} \times (\hat{V}_j - \bar{V}_j) + P_i(\hat{a}) \quad (3.57)$$

Index i refers to a wavelength of image pixel i. components of vector V, denoted as V_j, determine vertical columns of all gases, which has absorption lines in the selected spectral range. Factors of vertical distribution V_j and polynomial coefficients of vector a using as reduction parameters. Also, the atmosphere temperature profile is used as a parameter of smoothing.

Calculation parameters are determined as a minimum (on least squares method) between measured radiation (ln I^{obs}_i) and modeled radiation

(In $I^{mod}{}_i$), so by the length of vector RES (in particular, its components RES_i for every spectral measurement i).

For determination of the content of carbon dioxide using average on observation depth dry air molar fraction of this gas, measured at the same time with oxygen. So average on observation depth carbon dioxide concentration XCO_2 can be calculated as:

$$XCO_2 = \frac{CO_2^{col}}{(O_2^{col}/O_2^{mf})} \tag{3.58}$$

where $CO^{col}{}_2$ is received absolute content of CO_2 (molecules on cm^2), $O^{col}{}_2$ the same for O_2, $O^{mf}{}_2$ – expected (averaged on observation depth) oxygen molar fraction reduced to dry air (3,2095).

This is the way to calculate the tropospheric concentration of gas from radiation intensity in the frame of its absorption band. A method in details is described in [36].

The similar way we can use for determination of other gases concentration, for example, methane. But the concentration of carbon dioxide instead oxygen can be utilized in this algorithm [37].

$$XCH_4 = \frac{CH_4^{col}}{(CO_2^{col}/CO_2^{mf})} \tag{3.59}$$

where $CH^{col}{}_4$ is received absolute content of CH_4 (molecules on cm^2), $CO^{col}{}_2$ the same for CO_2, $CO^{mf}{}_2$ – expected (averaged on observation depth) carbon dioxide molar fraction reduced to dry air. For $CO^{mf}{}_2$ using constant 370 ppm.

The described WFM-DOAS algorithm was used for processing of observations of AIRS & AMSU (EOS satellite), SCIAMACHY (ENVISAT satellite) and GOSAT (Ibuki satellite) sensors, the distribution of atmospheric GHG concentrations was calculated.

3.4.2 Satellite Tools for Plant Productivity and Carbon Stock Assessment

Crop productivity control based on the possibility of the photosynthesis measurements. Carbon is accumulating in terrestrial ecosystems by photosynthesis and emitting as CO_2 through autotrophic and heterotrophic respiration [38]. GPP (Gross Primary Production) is the total carbon assimilated by vegetation through photosynthesis.

Net Primary Production (NPP) is a residue of detent carbon after accounting of autotrophic respiration. A significant part of detent carbon can be accumulated in living crops and organic soils, and emission this carbon to the atmosphere as CO_2 or methane can influencing to climate [38].

Satellite instruments collect parameters distributions describing crop activity. In particular, the data from the Collection 5 of the MOD17A2 of MODIS satellite sensor can be used for vegetation analysis. The MOD17A2 is an 8-day summation of the GPP and photosynthetic production (PsnNet). Annual GPP and NPP for MOD17A determined as [39]:

$$PsnNet = GPP - R_{ml} - R_{mr} \qquad (3.60)$$

where R_{ml} and R_{mr} respiration of leafs and roots.

Annual *NPP* can be calculated as [39]:

$$NPP = \sum_{i=1}^{365} PsnNet - (R_{mo} + R_g) = X_j \qquad (3.61)$$

where R_{mo} respiration of all part of plant excluding roots and leafs, R_g – respiration of growth (Zhao et al, 2005), X_j – observed *NPP* in the year j.

Model MOD17 based on three sources of input data. For every pixel land cover data provided by MOD12Q1 product. Meteorological data provided by DAO (Data Assimilation Office) model. Data on the spectral reflectance of surface FPAR i LAI provided from MOD15A2 product. Uncertainties of MOD12Q1, DAO, MOD15A2, as well as errors of the algorithm (3.7)–(3.48) may reflect to the correctness of output data MOD17.

Reliability of MOD12Q1 may be assessed in the interval [0.7–0.8] for the majority of land cover classes [40]. Moreover, meteorological data from DAO is not direct measured data, but reanalysis data. This type of data has both systematic and random errors [33]. These errors seriously reflect on the correctness of NPP calculation [40].

Algorithms MOD17, MOD15A2 are sensitive to the accuracy of input meteorological data (temperature and precipitation), both absolute values and distribution. Regional monitoring and control of vegetation productivity using satellite tools are effective and useful, but require additional instruments to uncertainty control and errors reducing.

Data on vegetation productivity is necessary for models of carbon balance, and for for climate change control on regional scale [38, 39]. Many significant climate-related risks can be assessed using vegetation production data: from agricultural and food security to landscape fires. For example, fire risk dynamics may be assessed using the forecasting of vegetation productivity [41].

In this context, the task of uncertainty control of vegetation productivity estimation is an urgent problem for decision making in GHG emission control, environmental security, and climate adaptation planning.

3.4.3 Satellite Tools for Uncertainty Assessment in Crops Productivity

Annual sums of monthly productivity and annual productivity estimations, obtained from satellite observations using the same methodology, could be compared to estimate relative uncertainty of productivity assessments. Time-series of these assessments for various regions allow detect key trends of errors, analyze components of uncertainty, connected with variations of local climate, land-use parameters.

For separate region m in month i, during year j from interval k, an error of productivity estimation U_m can be determined as the difference between observed annual productivity X_j and calculated monthly productivity x_{ij} according to:

$$U_m = \frac{\sum_j \frac{1}{k} ((\sum_i x_{ij}) - X_j)}{< x_{ij} >}. \tag{3.62}$$

In this case, a total uncertainty of productivity assessment for all regions m of study area M may be assessed as:

$$U_{tot} = \sqrt{\sum_m U_m^2}. \tag{3.63}$$

This algorithm has been utilized for analysis of productivity and calculation of uncertainty of vegetation of the Western part of Ukraine during observation period 2000–2012. The average regional error of productivity estimation was assessed according to [39].

Data shows that the average annual error is between 15,7 and 18,2%. In separate regions, it varied from 10,8% (regions with mainly homogeneous land-use structure, and low anthropogenic component), to 22,4% (landscapes and ecosystems with high productivity, high anthropogenic load, high heterogeneity of land covers).

Further consider uncertainty as for the aggregate of uncertainties, which connected with assessing the productivity. Thus, will assume that vectors of the true state of systems x at the moment of observation k ($j, j \in k$) are determined according to [31]:

$$x_k = R_k x_{k-1} + A_k u_k + w_k, \qquad (3.64)$$

where R_k – «system evolution matrix», e.g. formalized description of long-term managing impacts to productivity distributions (in our case – functional of climate processes) forming impacts to vector x_{k-1} at the moment k-1; A_k – «operating matrix», e.g. induced loads to system, influencing to productivity (such as land-use and land cover distribution), measured as impacts u_k to vector x; w_k – random process determining "noises" distribution, e.g. sensors, algorithms errors etc. In this case, an uncertainty can be determined as:

$$\delta x = \delta \hat{x} \frac{\partial R}{\partial t} + \delta \hat{x} \partial A + \delta w, \qquad (3.65)$$

where $\hat{x}$ – ensemble average values of the measured parameter, t – time.

Formalized description of long-term managing impacts to productivity distributions will consider as functional:

$$R = f(s) \cdot F(T, W, R) + \Delta, \qquad (3.66)$$

where $f(s)$ – function of density of vegetation (aggregate parameter of vegetation rarefaction, or degradation of crops related to business or to the crown branching), $F(T, W, R)$ – productivity functional, in general case depends of temperature, soil moisture and solar radiation (practically might be determined by remote observations), Δ – uncertainty coefficient [3].

If we assume the function of the density of vegetation the smooth homogeneous algebraic function, its behavior may be uniquely described using spectral vegetation indices. To describe operating impacts we can without generalization use the following form of equation [3]:

$$\delta \hat{x} \frac{\partial R}{\partial t} \rightarrow \frac{\sum_i (x_i - \hat{x}_i) \sum_j (\delta T_j - \delta \hat{T}_j)}{\sqrt{(\sum_i x_i - \hat{x}_i)^2 (\sum_j \delta T_j - \delta \hat{T}_j)^2}}$$

$$\times \frac{\partial (1 - \frac{\frac{1}{N}\sum_{n=1}^N T_n}{T^{\max}})(1 - \frac{1}{T^{\max} - \frac{1}{N}\sum_n T^{\max}})}{\partial t}, \qquad (3.67)$$

where T_n – air temperature during the observation period; T^{max} – maximal air temperature during the observation period; δT – «reduced temperature», $\delta T = (1 - \frac{\frac{1}{N}\sum_{n=1}^N T_n}{T^{\max}})(1 - \frac{1}{T^{\max} - \frac{1}{N}\sum_{n=1}^N T^{\max}})$. The parameter «reduced temperature» is an optimal correlator of parameters of ecosystems vulnerability and risks, for example, productivity and climatic parameters.

≪Reduced temperature≫ was calculated basing on the analysis of long-term correlative relationships in multivariate parameters distributions using elliptic kopulas [3]. Thus we approximate climatic impacts by changes of temperature. This is not absolutely correct, but measurements of temperature are the most reliable measurement among all parameters.

Induced loads to the ecosystem we will descry as impacts from land-use and land cover changes. Corresponding uncertainties connected with not correct classification and statistics (annual and seasonal), as well as significant variations of anthropogenic load to land on a local scale.

This type of uncertainty (uncertainty connected with the inaccuracy of classification of the observed system) we will through relative reliability of aggregated data. The ≪agreement index of suitability≫ (3.2) [27] may be used for it as:

$$\delta\hat{x}\partial A = \frac{\sum_i(x_i - \hat{x}_i)\sum_j(x_j^{norm} - \hat{x}_j^{norm})}{\sqrt{(\sum_i x_i - \hat{x}_i)^2(\sum_j x_j^{norm} - \hat{x}_j^{norm})^2}}$$

$$\times \frac{1}{n}\sqrt{\sum_{p,q}(\frac{x_p - x_p^{\min}}{x_p^{\max} - x_p^{\min}})(\frac{x_q - x_q^{\min}}{x_q^{\max} - x_q^{\min}})^2}, \qquad (3.68)$$

where q – study area sites; p – sets of statistical data (for example, field calibration measurements, statistics, cartography); x – measured value of the investigated parameter; x^{norm} – ≪normalized agreement index of suitability≫, $x_{norm} = \left(\sum_{p,q}(\frac{x_p - x_p^{\min}}{x_p^{\max} - x_p^{\min}})(\frac{x_q - x_q^{\min}}{x_q^{\max} - x_q^{\min}})^2\right)^{1/2}$. Using this approach we can estimate a mutual consistency of data distributions, most correctly combine disparate data, and so to minimize errors.

Distribution of uncertainties connected with methodical errors (aleatoric uncertainty) can be described as the process of measurements of system state vectors, e.g. as variations of random process w_k determining a set of sensor and processing errors. It can be assessed as:

$$\delta w = \sqrt{\sum_i^n (\Delta x_i \frac{\partial F}{\partial x})^2 + \nu'}, \qquad (3.69)$$

where $F(x)$ – calculated value on the sequence of measured parameters x $(x_1, x_2, \ldots, x_n)$ e.g. $F = F(x_1, x_2, \ldots, x_n)$; ν' – unknown error, which we interpret as inherent uncertainty. Component $\frac{\partial F}{\partial x}$ and so $\sqrt{\sum_i^n(\Delta x_i \frac{\partial F}{\partial x})^2}$

can be calculated using assessments of the accuracy of MOD17 FPAR i MOD15A2, according (Zhao et al., 2005, Wang, et al., 2001). Using data and results [3], we can estimate this component in diapason [0,045; 0,065].

Therefore, integrated form of the equation for assessment of components of uncertainty in the task of vegetation productivity assessment could be presented as [3]:

$$
\frac{\sum_j \frac{1}{k}((\sum_i x_{ij}) - X_j)}{<x_{ij}>} = \frac{\sum_i (x_i - \hat{x}_i) \sum_j (\delta T_j - \delta \hat{T}_j)}{\sqrt{(\sum_i x_i - \hat{x}_i)^2 (\sum_j \delta T_j - \delta \hat{T}_j)^2}}
$$

$$
\times \frac{\partial \left(1 - \frac{\frac{1}{N}\sum_{n=1}^N T_n}{T^{max}}\right)\left(1 - \frac{1}{T^{max} - \frac{1}{N}\sum_{n=1}^N T^{max}}\right)}{\partial t}
$$

$$
+ \frac{\sum_i (x_i - \hat{x}_i) \sum_j (x_j^{norm} - \hat{x}_j^{norm})}{\sqrt{(\sum_i x_i - \hat{x}_i)^2 (\sum_j x_j^{norm} - \hat{x}_j^{norm})^2}}
$$

$$
\times \frac{1}{n}\sqrt{\sum_{p,q}\left(\frac{x_p - x_p^{min}}{x_p^{max} - x_p^{min}}\right)\left(\frac{x_q - x_q^{min}}{x_q^{max} - x_q^{min}}\right)^2}
$$

$$
+ \sqrt{\sum_i^n \left(\Delta x_i \frac{\partial F}{\partial x}\right)^2} + \nu' \qquad (3.70)
$$

Separate components of uncertainty could be analyzed using this equation in the task of assessment of vegetation productivity by satellite observation data.

It should be noted, that this equation may be reduced for calculation of carbon balance up to constant (Shvidenko, et al., 2010), as an algorithm for determination of standard error $m^2{}_y$ of carbon flux $Y = f(X_i)$, determined as random distribution function X_i, $i = 1, 2, \ldots k$:

$$
m_y^2 = \sum_{i=1}^k \left(\frac{dY}{dX_i} m_i\right)^2 + 2\sum_{i>j}\left(\frac{dY}{dX_i}\right)\left(\frac{dY}{dX_j}\right) r_{ij} m_{X_i} m_{X_j} \qquad (3.71)
$$

where r_{ij} – coefficients of correlations between distributions X_i and X_j.

This algorithm allows estimate errors of carbon fluxes estimations in all components of the studied system using all available data. It is important, that it is possible to incorporate into whole approach data of modeling, in-field ground measurements, and satellite observations.

As the calculations show [3], estimated the uncertainty of assessment of vegetation productivity by satellite data is [0.138; 0.152].

Constant component (aleatoric uncertainty) of uncertainty is errors of the sensor, processing of data (calculation of indexes, estimation of biophysical parameters from reflection, etc.) and monitoring methods (non-optimal time of observation, incorrect data integration, etc.). This uncertainty component is about [0.045; 0.065]. This constant component cannot be essentially reduced by the data user. Varied component, which corresponds to regional variations of climatic parameters, lies in the interval [0.02; 0.035]. This component may be reduced with an account of correct regional climate trends. Significant uncertainty component – [0.053; .074] is caused by incorrect classification of land cover and imperfect analysis of local land-use structure. This type of uncertainty (epistemic) is connected with a lack of information about system studied, and requires more comprehensive models to reduce it. Besides ground calibration data and observation verification are required too. Besides, there is an uncertainty component about [0.01; 0.005] of unknown origin, which we identify as inherent uncertainty.

Therefore, it can be argued, that integration of Earth systems modeling, analysis of satellite data, and processing of ground measurements for calibration and verification allow to reduce 55–65% of error of vegetation productivity estimation.

3.5 Approach to Emission Assessment and Control

Basing on measured atmospheric GHG concentrations it is possible to calculate corresponding GHG emissions.

The simplest way to calculate aggregate annual GHG emission by its concentrations, according to [20, 42] is:

$$E_i = \sum_i \kappa_i \frac{(C_{current} - C_{previous})M_i}{C_{0i}} / t \qquad (3.72)$$

here $C_{current}$ – current mean annual atmospheric concentration of gas i over certain observed territory; $C_{previous}$ – previous current mean annual atmospheric concentration of gas i over certain observed territory; C_{0i} – mean annual atmospheric concentration of gas i over same territory during pre-industrial era (according to geological data and historical records); M_i – historical mass of estimated GHG in the atmosphere (according to geological data and historical records) [43]. For carbon dioxide and methane C_0 is 275 and 700 ppmv; and M_i – 220,89 × 10^{10} tons for CO_2 and 408,94 × 10^{10} tons for CH_4 [44] for whole Earth atmosphere. Coefficient κ_i

determining with an account of a lifetime of GHG and its mass at the moment of estimation (for short-term forecasting it possible to assume $\kappa_i = 1$ for all GHG); t – averaging period (in years).

This algorithm allows to control national GHG emissions by independent tools (satellite observations). Analysis of identified differences (Figure 3.6) is a base for decision making if the field of national emission management and environmental protection policy.

In particular, the detected difference of assessments could indicate inadequateness the national inventory system during the observation period. This data also should focus attention on importance ecological processes. Input these processes into GHG balance could be 19–23% and it should adequately be reflected in emission report data.

In the case of described distributions (3.2)–(3.4) we can present a modified with Equation (3.9) form of uncertainty:

$$U_n = \left(|E^1_{tot,n} - E^2_{tot,n}|\right) / \left(\frac{\sigma_i \sum_{i,m} \left(\frac{|E^1_{tot,n} - E^2_{tot,n}|}{2} - \sum_i E^1_{tot,n} \right)}{CoVar_m} \right) \quad (3.73)$$

Calculations show that assessments based on ecological models and observation data are more correct and accurate on all scales than statistics [3].

3.5.1 Stochastic Tools for Decision Making in Carbon Emissions Control

A fundamentally different approach to decision making under uncertainty in the field of environmental management was proposed by [45], basing on robust cost-effective and environmentally safe carbon trading economic instruments for GHG emissions control.

This approach based on the stochastic model of robust emission trading.

Proposed integrated robust multi-agent emission trading model seeks the goal that all the parties participating in emissions trading jointly achieve individual emission targets in a cost-efficient way under safety constraints by investing in emissions abatement, uncertainty reduction and by redistributing the emission permits through trading.

Emission trading affects and is being affected by emission uncertainties. In addition to asymmetric information, it implicitly contributes to the reduction of other uncertainties for the two main reasons. First, trading is likely to lead towards more scrutinized emissions inventory compilation rules. Second, verifiability of trades requires that the reported emissions plus uncertainties are below the cap; therefore, trading creates incentives for parties to invest in uncertainty reduction prior to compliance. Proposed trading model exploits this phenomenon.

Different uncertainties differently affect emission trading causing market crashes and instabilities similar to financial markets. To limit the role of uncertainties, advocates of regulated trades argue in favor of uncertainty constraints distinguishing sources by their uncertainty levels [46]. Market regulators may set restrictions on source category to be included in trading. The trading scheme may demand a Party to set source-specific targets depending on the level of uncertainty.

A model which includes uncertainty and risk-adjusted regulations into emission trading schemes is proposed following [47]. The model explores conditions of market's stability towards uncertainties by imposing appropriate safety constraints to control the level of admissible uncertainty which would guarantee cost efficiency of trades and safety levels of emission reduction targets.

These types of safety constraints are typical for pollution control, financial applications, stability regulations in the insurance industry and catastrophic risks management [45]. In a sense, these constraints work as a probabilistic discounting mechanism which discounts the reported emissions to detectable levels overshooting uncertainty within a specified safety level, i.e., a portion of detectable emission changes.

The approach can be formally presented as follow [45]. Let denote the least costs $f_i(y_i)$ for party i to comply with imposed targets with given permits y_i and the target K_i. Formally, this function is defined by Equations (3.) – (3.). Denote the variability of reported emission x_i as a random variable $\xi_i(x_i, \omega_i)$, where ω_i is a vector of all uncertainties (scenarios) affecting emissions of party i. A random variable $\xi_i(x_i, \omega_i)$ depends, in general, on reported emissions x_i. The uncertainty ξ_i can be reduced by investments in monitoring systems. For this purpose, the variable u_i associated with monitoring and other technologies that may control the variability of emissions within the desirable safety level Q_i is introduced.

The individual optimization problem can be written as minimization of a function $f_i(y_i)$ defined as the minimum of risk-adjusted expected emission reduction costs $c_i(x_i, \omega_i)$ and uncertainty reduction costs $d_i(x_i, \omega_i)$ for a given permit y_i to be defined through a dynamic trading process:

$$f_i(y_i) = \min_{x_i u_i} E[c_i(x_i, \omega) + d_i(u_i, \omega)] \tag{3.74}$$

under quantile-based environmental safety constraints:

$$P[x_i + \xi(x_i, \omega_i) \leq K_i + y_i] \geq Q_i \tag{3.75}$$

for all parties i. Here Q_i denotes a safety level that ensures the probability that all potential emissions x_i and uncertainties $\xi_i(x_i, \omega_i)$ do not exceed the emission target K_i adjusted by permits y_i. The safety level Q_i is imposed by a regulatory agency to ensure the robust performance of the market. Uncertainties of cost functions c_i and d_i may be due to market performance, production shocks, and technological uncertainties, which are unknown in advance.

Safety constraints (3.5) are well known in financial applications as Value-at-Risk risk indicator. They are used for the safety regulation of insurance companies, the reliability of engineering structures, and catastrophic risk management [45].

Safety constraints (3.5) can be also written in the following form that in the case of analytically tractable distributions is reduced to deterministic nonlinear constraints (3.6).

Let define quantile $z_i(x_i)$ as the minimal z such that $P[\xi_i(x_i, \omega_i) \leq z] \geq Q_i$.

Then the following equivalent constraints can be substituted for the safety constraint (3.5):

$$x_i + u_i \leq K_i + y_i, u_i \leq z_i(x_i) \tag{3.76}$$

These equations show that reported emissions must undershoot targets K_i adjusted by uncertainties of emissions u_i and permits y_i. Equation (3.1) shows that safety constraints induce risk-related upper bounds $z_i(x_i)$ on uncertainty intervals dependent on the reported emission level x_i. Therefore, it allows introducing risk-based undershooting of emission targets defined by "critical" quantile $z_i(x_i)$. Functions $f_i(y_i)$ define costs of permits y_i for parties $i = 1,...,n$.

In [45] solution of this task is proposed for different cases. For our study, it is important that in this framework the task of uncertainty reducing for decision making optimization is the task of uncertainty u_i control in GHG emission x_i inventory. Further, we demonstrate as it can be done using satellite tools.

3.6 Results of Multiparametric Modeling of Methane Cycle in Temperate Wetlands in View of Regional Climate Change Using Satellite Data

The proposed approach to modeling GHG emissions allows calculating its individual components. In particular, we can consider the actual problem of methane emissions from temperate wetland ecosystems.

The basis of this review is the methodological approach described in [3, 5] and in Sections 3.3, 3.4, 3.5 of this chapter.

The National Report [7] recorded a decrease in total methane emissions (Figure 3.3), which, however, is accompanied by an increase of the concentration of methane in the atmosphere over the entire territory of Ukraine during the investigated period (Figure 3.4). Thus, there is a contradiction between the parameters calculated according to the deterministic method [20] and obtained with the independent observation instruments and stochastic approaches [3]. Under the uncertainty of 150% level, this requires an understanding of the factors and drivers of the ongoing processes, as well as significant improvement of the models.

As has been shown in Section 3, on the basis of consideration of the global model of atmospheric balance of CH4 (in particular, the CLM4Me methane production model, which is a component of the geosynthetic model CESM1 included in the global climate model CCSM4) an approved flood integrated coefficient obtained from regional observations can be proposed. Also, in Section 3.4 it is shown that in the framework of the general approach, there is the possibility to refine global models on the local observations and models, as well as the climatic scenarios, can be embedded into local adaptation decision making systems.

So, with the purpose to correction certain indicators of models at the local level, partial improvement of the model approach to the evaluation of individual components of wetlands methane emissions was proposed. In

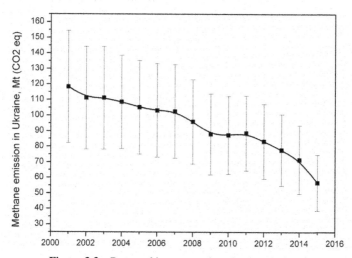

Figure 3.3 Reported inter-annual methane emissions.

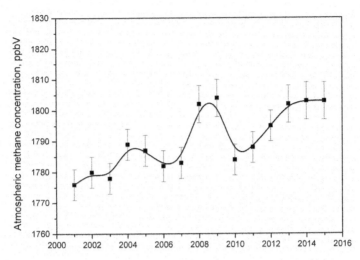

Figure 3.4 Detected inter-annual atmosphere methane concentration.

particular, a model of wetlands was proposed, aimed at the assessment of the methane emissions, depending on changes of climatic parameters.

Model variables can be controlled using satellite tools have been identified. For the e, an of key variables was conducted.

Analysis of the main trends of changes of specific indicators of key variables has been conducted directed to analysis of the dynamics of emissions from the wetlands of the North-Western part of Ukraine. In particular, the methane concentrations during the observation period using the data from the AIRS and AMSU/Aqua satellites were controlled. Using the algorithms for converting the spectral intensity values to concentrations, detailed spatial and temporal parameters of methane distribution in the atmosphere were estimated (Figure 3.4). These data were used for calibration and verification of the model (Figure 3.5).

Basing on the proposed local model, critical variables were identified, which were further controlled by observations and measurements to solve the problem of estimating methane production in wetlands in the context of climate change. These were surface and near-surface air temperatures, soil moisture, and saturation of the near-surface layer, precipitation, water level (flood), available biomass and evapotranspiration.

The control of the specified variables was carried out both on the annual and on the monthly distribution. The distribution of precipitation (Figure 3.6), temperature (Figure 3.7), evapotranspiration and vegetation index (Figure 3.8) corresponds to the distribution of atmospheric methane

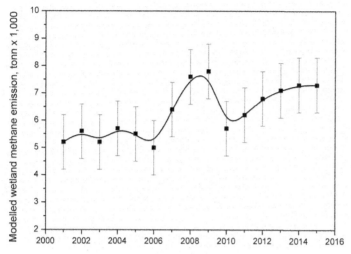

Figure 3.5   Modeled inter-annual methane emissions.

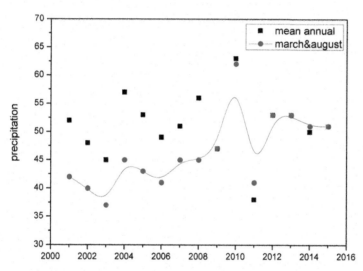

Figure 3.6   Detected inter-annual of mean and selected month distribution of precipitation over wetlands, mm.

concentrations over the wetlands during the observation period (Figure 3.4) and to the emission models (Figure 3.5).

This means that the proposed model adequately reflects key processes that affect methane production and correctly identifies variables that allow control the methane emissions from temperate wetlands.

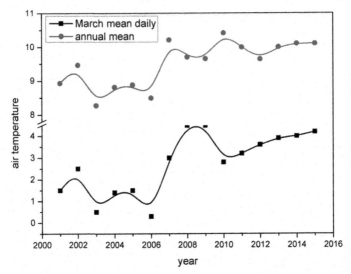

Figure 3.7   Detected inter-annual of mean and selected month distribution of air temperature over wetlands, °C.

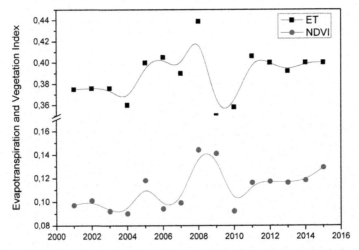

Figure 3.8   Detected inter-annual evapotranspiration and vegetation index.

According to the model calculation, verified the observations and measurements, the average methane production from Ukrainian temperate wetlands can be estimated in interval 0,2–0,8 gC/m^{2}day during the observation period.

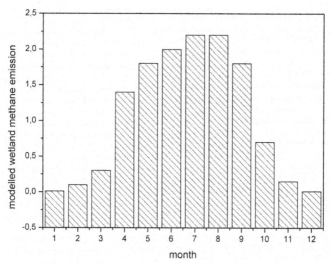

Figure 3.9 Modeled intra-annual atmosphere methane emissions, ton x 1,000.

In addition, the comparison of data obtained from satellites with meteorological measurements made it possible to determine the observation methodology: to determine the observation periods critical for methane emissions in the wetland systems of Ukraine. The analysis of the distribution of meteorological data on wetland areas during vegetation periods associated with the activity of the methane cycle allowed to determine the periods March-April and August-September (Figure 3.9), when the emission gradient is the largest.

According to the results of this study, improvement of the monitoring methodology, aimed to methane cycle indicators assessment, may be proposed. In particular, the use of modified vegetation spectral indices in the framework of proposed methane balance models as well as the observation during certain periods should reduce the inherent uncertainty of this task.

3.7 Conclusions

In this chapter, the advantages of applied mathematics based approaches for ecology, environmental security, climate regional studies oriented to change adaptation, and for remote sensing technologies development is demonstrated.

Different approaches to anthropogenic and natural emissions inventory and analysis are described and discussed. Possible algorithms of uncertainty control for discussed approaches are proposed.

The task of analysis of climate change at the regional and local levels, in particular, the adaptation of the economy to the escalation of negative phenomena (including hazards), requires the assessment of changes of GHG emissions and the dynamics of their atmospheric balance, both at the global level and over the separate areas. In particular, methane requires a serious attention as the second largest greenhouse gas after carbon dioxide. At the same time parameters of methane emissions and absorption, distributions and its dynamics, especially its radiation impact on the energy of the atmosphere, are much less studied. The general idea of this study is to demonstrate the capabilities of satellite observation tools used in the model approach to clarify individual methane emissions. Currently, the assessment of major sources of methane has significant uncertainties. The error of estimation (aleatoric uncertainty) is about 20–40%. At the same time, an error in the estimation of the main driving forces and emission parameters (epistemic uncertainty) is more than 150%. The National Greenhouse Gas Inventory Report also indicates approximately the following parameters. This situation requires the use of scientific tools to reduce uncertainties.

It is shown that a decisive role in uncertainty reducing in the tasks of GHG emission plays a correct analysis of the carbon cycle, and so – mathematical models of ecosystems. For determination of key model variables, measured parameters and indicators for satellite observations the GHG and carbon cycle models have been analyzed. Advanced model of methane emission was proposed. Uncertainties in carbon balance models are analyzed, and algorithms to reduce it are proposed.

Basing on the models described the satellite tool for GHG monitoring is defined and discussed. An algorithm for calculation of individual gas concentrations in the atmosphere from radiation intensity within the separate absorption bands is analyzed.

As the important component of GHG monitoring system, a satellite tool for plant productivity assessment is analyzed. A multicomponent stochastic model for uncertainty assessment in productivity analysis is proposed. As the separate, intercorrelated and multivariate components of uncertainty the instrumental errors, climate impact, and land-use change factors have been determined and analyzed. It was shown that quantitative assessments of these components in its temporal dynamics allow obtaining new knowledge about investigated tools, climatic parameters and ecosystem we study.

Algorithm for emission calculation from atmospheric GHG concentration is proposed. Approach and algorithm to analyze and reduce uncertainty in emission account are proposed and discussed.

In the framework of the analysis, the main global models of atmospheric CH_4 balance were considered, and the key sources of emissions at the global and regional levels were evaluated. It was noted that the reduction of the total methane emissions recorded in the National Report is accompanied by an increase of the concentration of methane in the atmosphere over the entire territory of Ukraine during the period of investigation. This requires an understanding of the factors and drivers of the ongoing processes, the substantial improvement of models and tools. The global models of the atmospheric balance of CH_4 were considered: in particular, the model of production of methane CLM4Me, which is a component of the geosynthetic biosphere model CESM1, which is part of the global climatic model CCSM4. In the framework of this model, an improved look at the integrated flood rate obtained from regional observations was proposed. It is shown that in the general approach, with the possibility of refining global models on local observations and models, climatic scenarios can be embedded in local adaptation decision making systems. In order to clarify certain model indicators at the local level, partial improvement of the model approach to the assessment of individual components of natural methane emissions was proposed. In particular, a model of wetlands was proposed, aimed at the determination of the emissions of methane, depending on changes in climatic indicators.

Conclusions on the reliability and correctness assessment of decisions based on the approach discussed are demonstrated. We can conclude that proposed approaches to account emission parameters, as well as both aleatoric and epistemic uncertainty in complex systems, is the more correct base for decision making than usual way based on deterministic approaches.

Acknowledgments

The authors are grateful to anonymous referees for constructive suggestions that resulted in important improvements to the chapter, to colleagues from the International Institute for Applied Systems Analysis (IIASA), from the American Statistical Association (ASA), American Meteorological Society (AMS), and from the International Association for Promoting Geoethics (IAPG) for their critical and constructive comments and suggestions. The authors express thanks to the Science and Technology Center in Ukraine (STCU) for partial support of this study in the framework of research project #6165 "Information and Technological Support for Greenhouse Effect Impact Assessment on Regional Climate using Remote Sensing."

References

[1] N. Fenton, M. Neil, 'Risk assessment and decision analysis with Bayesian networks', CRC Press, Boca Raton, FL, 2012.

[2] Y. V. Kostyuchenko, 'On the methodology of satellite data utilization in multi-modeling approach for socio-ecological risks assessment tasks: A problem formulation', IJMEMS, 3(3.), 1–8, 2018.

[3] Y. V. Kostyuchenko, D. Movchan, I. Artemenko, & I. Kopachevsky, 'Stochastic Approach to Uncertainty Control in Multiphysics Systems: Modeling of Carbon Balance and Analysis of GHG Emissions Using Satellite Tools', *Mathematical Concepts and Applications in Mechanical Engineering and Mechatronics*, ed. by Mangey Ram and J. Paulo Davim, IGI Global, USA, 2017, pp. 350–378.

[4] M. Cao, S. Marshall, & K. Gregson, 'Global carbon exchange and methane emissions from natural wetlands: Application of a process based model'. *Journal of Geophysical Research: Atmospheres, 101*(D9), 14399–14414, 1996.

[5] M. A. Popov, S. A. Stankevich, Yu. V. Kostyuchenko, A. A. Kozlova, 'Analysis of Local Climate Variations Using Correlation between Satellite Measurements of Methane Emission and Temperature Trends within Physiographic Regions of Ukraine', IJMEMS, 4(3.), 276–288, 2019.

[6] K. L. Denman, G. Brasseur, A. Chidthaisong, P. Ciais, P. M. Cox, R. E. Dickinson, D. A. Hauglustaine, C. Heinze, E. Holland, E. Jacob, U. Lohmann, S. Rmachandran, P. L. da Silva Dias, S. C. Wofsy, and X. Zhang, 'Couplings between changes in the climate system and biogeochemistry, in: Climate Change 2007: The physical science basis. Contribution of Working Group I to the Fourth Assessment Report of the Intergovernmental Panel on Climate Change', edited by: Solomon, S., Qin, D., Manning, M., Chen, Z., Marquis, M., Averyt, K. B., Tignor, M., and Miller, H. L., Cambridge University Press, Cambridge, United Kingdom and New York, NY, USA, 512 pp., 2007.

[7] VI National Communication on the Climate Change of Ukraine. (3.014). Kiev: Ministry of Environment and Natural Resources of Ukraine, 323 p. http://www.seia.gov.ua/seia/document/638134.

[8] M. A. K. Khalil, and M. J. Shearer, 'Sources of methane: An overview', Atmospheric methane: Its role in the global environment, edited by: Khalil, M. A. K., Springer, New York, NY, 98–111, 2000.

[9] P. Bousquet, P. Ciais, J. B. Miller, E. J. Dlugokencky, D. A. Hauglustaine, C. Prigent, G. R. Van der Werf, P. Peylin, E.-G. Brunke, C. Carouge, R. L. Langenfelds, J. Lathiere, F. Papa, M. Ramonet, M. Schmidt, L. P. Steele, S. C. Tyler, and J. White, 'Contribution of anthropogenic and natural sources to atmospheric methane variability', Nature, 443, 439–443, 2006.

[10] E. A. G. Schuur, J. Bockheim, J. G. Canadell, E. Euskirchen, C. B. Field, S. V. Goryachkin, S. Hagemann, P. Kuhry, P. M. Lafleur, H. Lee, G. Mazhitova, F. E. Nelson, A. Rinke, V. E. Romanovsky, N. Shiklomanov, C. Tarnocai, S. Venevsky, J. G. Vogel, and S. A. Zimov, 'Vulnerability of permafrost carbon to climate change: Implications for the global carbon cycle', Bioscience, 58, 701–714, 2008.

[11] G. M. MacDonald, D. W. Beilman, K. V. Kremenetski, Y. W. Sheng, L. C. Smith, and A. A. Velichko, 'Rapid early development of circum-arctic peatlands and atmospheric CH_4 and CO_2 variations', Science, 314, 285–288, 2006.

[12] R. Wania, I. Ross, and I. C. Prentice, 'Integrating peatlands and permafrost into a dynamic global vegetation model: 2. Evaluation and sensitivity of vegetation and carbon cycle processes', Global Biogeochem. Cy., 23, Gb3015, 2009.

[13] K. M. Walter, , L. C. Smith, and F. S. Chapin, 'Methane bubbling from northern lakes: present and future contributions to the global methane budget', Philos. T. R. Soc., 365, 1657–1676, 2007.

[14] R. Segers, C. Rappoldt, and P. A. Leffelaar, 'Modeling methane Fluxes in wetlands with gas-transporting plants 2. Soil layer scale', J. Geophys. Res.-Atmos., 106, 3529–3540, 2001.

[15] K. A. Smith, T. Ball, F. Conen, K. E. Dobbie, J. Massheder, and A. Rey, 'Exchange of greenhouse gases between soil and atmosphere: interactions of soil physical factors and biological processes', Eur. J. Soil Sci., 54, 779–791, 2003.

[16] T. R. Christensen, T. R. Johansson, H. J. Akerman, M. Mastepanov, N. Malmer, T. Friborg, P. Crill, and B. H. Svensson, 'Thawing sub-arctic permafrost: Effects on vegetation and methane emissions', Geophys. Res. Lett., 31(3.), L04501, 2004.

[17] J. O. Kaplan, 'Wetlands at the Last Glacial Maximum: Distribution and methane emissions', Geophys. Res. Lett., 29, 1079, 2002.

[18] F. Keppler, J. T. G. Hamilton, M. Brass, and T. Rockmann, 'Methane emissions from terrestrial plants under aerobic conditions', Nature, 439, 187–191, 2006.

[19] A. A. Bloom, P. I. Palmer, A. Fraser, D. S. Reay, and C. Frankenberg, 'Large-Scale Controls of Methanogenesis Inferred from Methane and Gravity Spaceborne Data', Science, 327, 322–325, 2010.

[20] 'Revised IPCC Guidelines for National Greenhouse Gas Inventories: Reporting Instructions', The Workbook, Reference Manual, Vol.1–3, IPCC, 1997.

[21] T. White, M. Jonas, Z. Nahorski, S. Nilsson, 'Greenhouse gas inventories. Dealing with uncertainties', Springer, Heidelberg, 343p., 2011.

[22] R. A. Bun, D. V. S. achuk, 'Modeling of global and regional carbon cycles in biosphere' Problems of operating research and informatics, #2, 141–148, 1997.

[23] P. R. Gent, G. Danabasoglu, L. J. Donner, M. M. Holland, E. C. Hunke, S. R. Jayne, D. M. Lawrence, R. B. Neale, P. J. Rasch, M. Vertenstein, P. H. Worley, Z.-L. Yang, and M. Zhang, 'The community climate system model version 4', *Journal of Climate*, 24(3.9), 4973–4991, 2011.

[24] D. M. Lawrence, K. W. Oleson, M. G. Flanner, P. E. Thornton, S. C. Swenson, P. J. Lawrence, X. Zeng, Z. -L. Yang, S. Levis, K. Sakaguchi, G. B. Bonan, and A. G. Slate., 'Parameterization improvements and functional and structural advances in version 4 of the Community Land Model', *Journal of Advances in Modeling Earth Systems*, 3(3.), 2011.

[25] J. R. M. Arah, & K. D. Stephen, 'A model of the processes leading to methane emission from peatland', *Atmospheric Environment*, 32(3.9), 3257–3264, 1998.

[26] H. Lieth, 'Primary production of the major vegetation units of the world', *Primary productivity of the biosphere,* pp. 203–215, Springer, Berlin, Heidelberg, 1975.

[27] A. Shvidenko, D. Schepaschenko, I. McCallum, S. Nilsson, 'Can the uncertainty of full carbon acconting of forest ecosystem be made acceptable to policymakers?' Climate Change, 103, pp. 137–157, 2010.

[28] J. M. Boardman, R. M. Vogt, 'Homotopy invariant algebraic structures on topological spaces', Springer-Verlag, Berlin, 407p., 1973.

[29] H. Raiffa, and R. Schlaifer, 'Applied statistical decision theory', Harward University Press, Boston. 356 pp., 1961.

[30] C. D. Lloyd 'Local models for spatial analysis', CRC Press, Taylor & Francis Group, 244 p., 2007.

[31] M. Aoki, 'Optimization of stochastic systems: topics in discrete-time systems', New York-London, 424p., 1967.

[32] K. Villez, M. Ruiz, G. Sin, J. Colomer, C. Rosen and, P. A. Vanrolleghem, 'Combining multiway principal component analysis (MPCA) and clustering for efficient data mining of historical data sets of SBR processes', Water Science & Technology, **57**, # 10, pp. 1659–1666, 2008.

[33] Yu. V. Kostyuchenko. 'Geostatistics and remote sensing for extremes forecasting and disaster risk multiscale analysis', in Numerical Methods for Reliability and Safety Assessment: Multiscale and Multiphysics Systems, S. Kadry and A. El Hami, (eds.), 404–423, Switzerland: Springer International Publishing, 2015.

[34] X. Wang, C. H. Bishop, 'A Comparison of Breeding and Ensemble Transform Kalman Filter Ensemble Forecast Schemes', Journal of the Atmospheric Sciences, vol. 60, pp. 1140–1158, 2003.

[35] J. -M. Lee, C. K. Yoo, S. W. Choi, P. A. Vanrolleghem, I. -B. Lee, 'Nonlinear process monitoring using kernel principal component analysis', Chemical Engineering Science, **59**, pp. 223–234, 2004.

[36] M. Buchwitz, R. de Beek, J. P. Burrows, H. Bovensmann, T. Warneke, J. Notholt, J. F. Meirink, A. P. H. Goede, P. Bergamaschi, S. Korner, M. Heimann, and A. Schulz, 'Atmospheric methane and carbon dioxide from SCIAMACHY satellite data: Initial comparison with chemistry and transport models', Atmos. Chem. Phys., 5, 941–962, 2005.

[37] O. Schneising, M. Buchwitz, J. P. Burrows, H. Bovensmann, P. Bergamaschi, and W. Peters, 'Three years of greenhouse gas column-averaged dry air mole fractions retrieved from satellite – Part 2: Methane', Atmos. Chem. Phys., 9, 443–465, 2009.

[38] M. Heimann and M. Reichstein, 'Terrestrial ecosystem carbon dynamics and climate feedbacks', *Nature, 451,* 289–292, 2008.

[39] D. Movchan, Yu. V. Kostyuchenko, 'Regional Dynamics of Terrestrial Vegetation Productivity and Climate Feedbacks for Territory of Ukraine', Int. J. of Geographical Information Science, Issue 6, pp. 889–912, 2015.

[40] M. Zhao, S. W. Running, & R. R. Nemani, 'Sensitivity of Moderate Resolution Imaging Spectroradiometer (MODIS) terrestrial primary production to the accuracy of meteorological reanalyses', *Journal of Geophysical Research, Vol. 111, G01002, 2006.*

[41] J. E. Janisch, M. E. Harmon, 'Successional changes in live and dead wood carbon stores: implications for net ecosystem productivity' Tree Phys., 22, pp. 77–90, 2002.

[42] J. Houghton, 'Climate change and sustainable energy', Met Office, Hadley Centre, FitzRoy Road, Exeter EX1 3P8, London UK, 208 p., 2005.

[43] D. Raynaud, J. M. Barnola, J. Chappellaz, T. Blunier, A. Indermuhle and B. Stauffer, 'The ice record of greenhouse gases: a view in the context of future changes', Quaternary Science Reviews, 19, 9–17, 2000.

[44] J. R. Petit, J. Jouzel, D. Raynaud, N. I. Barkov, J. M. Barnola, I. Basile, M. Bender, J. Chappellaz, J. Davis, G. Delaygue, M. Delmotte, V. M. Kotlyakov, M. Legrand, V. M. Lipenkov, C. Lorius, L. Pépin, C. Ritz, E. Saltzman, M. Stievenard, 'Climate and atmospheric history of the past 420,000 years from the Vostok ice core, Antarctica', Nature, 399, 429–436, 1999.

[45] T. Ermolieva, Y. Ermoliev, M. Jonas, M. Obersteiner, F. Wagner, W. Winiwarter, 'Uncertainty, cost-effectiveness and environmental safety of robust carbon trading: integrated approach', Climatic Change, Volume 124, Issue 3, pp 633–646, 2014.

[46] O. Godal, Y. Ermoliev, G. Klassen, M. Obersteiner, 'Carbon trading with imperfectly observable emissions', Environ Res Econ 25,151–169, 2003.

[47] T. Ermolieva, Y. Ermoliev, G. Fischer, M. Jonas, M. Makowski, 'Cost effective and environmentally safe emission trading under uncertainty', In: Marti K, Ermoliev Y, Makowski M (eds) Coping with uncertainty: Robust solutions. Springer-Verlag, Heidelberg, Germany, pp. 79–99. 2010.

4

An Intelligent Neuro Fuzzy System for Pattern Classification

Abdu Masanawa Sagir[1,*] and Saratha Sathasivam[2]

[1]Department of Basic Studies, Hassan Usman Katsina Polytechnic, Katsina, Nigeria
[2]School of Mathematical Sciences, Universiti Sains Malaysia, 11800 Pulau Pinang, Malaysia
E-mail: amsagir@yahoo.com; saratha@usm.my
*Corresponding Author

This research aims to develop a capable, intelligent system that can be used for classification purposes. This research describes a methodology for developing a smart system by utilising the application of an adaptive neuro-fuzzy inference system that can be used by physicians to accelerate the diagnosis process. The proposed intelligent classification system maximises the correctly classified data and minimises the number of incorrectly classified patterns. For robustness, the proposed method was tested with five different datasets, namely Hepatitis, SPECT Heart, Cleveland Heart, Diabetic Retinopathy Debrecen and BUPA Liver Disorders. Also, an attempt was made to specify the effectiveness of the performance metrics. The proposed method achieves superior performance when compared to related existing methods. This research has produced three major novelty techniques. Firstly, the Jacobian matrix was built up using an analytical derivation scheme and chain rule, which overcome the complexity for computation of the Jacobian matrix instead of finite difference schemes. Secondly, indexing unique membership functions in a row-wise vector using a vectorisation technique to obtain efficient classification performance results and increase the speed of convergence rates was introduced. Thirdly, an effective way to transform Jacobian matrix into a sparse Jacobian matrix was employed, which also contributed towards reducing memory consumption or storage space, improved execution time and accelerated the processing of data.

4.1 Introduction

Currently, most real-world classification systems are considered difficult in the developmental process, as they may involve numerous trade-off problems like computational cost, accuracy, storage space and execution time. They may be linear or non-linear, predictable or unpredictable. Developing an intelligent neuro-fuzzy system is of importance in almost all fields, especially in medical disease diagnosis, transportation, signal processing, telecommunication and engineering [1].

Neuro-fuzzy systems are multilayer connectionist networks that realize the essential elements and functions of traditional fuzzy logic decision systems. The performance of the system depends on the accuracy of the model. Therefore, it is of greatest importance to build a model which correctly reflects the behaviour of the system under consideration. The Sugeno type of fuzzy model was turned based on training data. It illustrates how natural language can be encoded and shows the relationship of entities in terms of IF-THEN rules.

Physicians make use of digital technologies to assist in diagnosis and give suggestions, as medical diagnoses are full of uncertainty. The best and efficient technique for dealing with change is applying soft computing techniques and incorporating fuzzy logic and neural networks, which are complementary to each other rather than fighting for system identification or recognition. This approach has the capability of becoming familiarized with patterns and adapt to themselves to handle a changing environment [2]. The essential contribution of neural networks is a methodology for identification, learning and adaptation, while the crucial input of fuzzy logic is a methodology for computing with words which can deal with ideas or knowledge and little details about data.

Classification techniques are being used in different fields of studies to quickly identify the type and group to which a tuple belongs. Classification is part of pattern recognition and aims to classify data (patterns) based on either prior knowledge or statistics and information extracted from the models. Grading is about predicting class labels using input data. In binary classification, there are two possible output classes [3].

4.2 Artificial Neural Networks Approach

Artificial Neural Network (ANN) is the most popular pattern classification approach. ANN may be traced back to as early as 1943 when McCulloch and Pitts wrote their landmark paper "A Logical Calculus of Ideas Immanent in Nervous Activity". Since then scientists, engineers, physicists and biologists have studied ways to mathematically model the human brain's ability to accept inputs from many completely different types of sensors. They can then analyse that data, decide on a course of action, trigger the response in its operational appendages and learn from the results of that response [4, 5].

ANN is a simplified mathematical model of the human brain. It is a parallel distributed processor with large numbers of connections; an information processing system that has specific performance characters like biological neural networks [6]. The main characteristic of the ANN-based classification approach is its ability to realize complex and non-linear input-output relationships. Artificial neural networks could be a type of statistical pattern classifier [7].

4.3 Statistical Approach

The statistical decision theoretic approach includes decision making in the presence of statistical knowledge, which provides some information where there is uncertainty. There are well-known decision rules such as Bayes decision, minimum error-rate classification and minimum likelihood. In general, most recognition models of statistical decision theoretic approach operate in two modes: training (learning) and classification (testing). In the training mode, the feature extraction module finds the appropriate features to represent the input patterns, and the classifier is trained to partition the feature space. In the classification mode, the trained classifier assigns the input pattern to one of the pattern classes under consideration based on the measured features [8].

4.4 Design of an Intelligent Neuro Fuzzy System

A typical Fuzzy model (4.1) was used and has the form [9]

$$IF\ x = A\ and\ y\ is\ B\ THEN\ z = f(x, y) \qquad (4.1)$$

where A and B are fuzzy sets in the rule antecedent part, while $z = px + qy + r = f(x, y)$ is a crisp function in the rule consequent part and p, q & r

are the optimal consequent parameters. Usually $f(x,y)$ is a polynomial in the input variables x and y.

4.5 Hybrid Learning Algorithm of Proposed Method

When developing the proposed system, parameters are calculated by the hybrid learning technique (forward and backwards pass) based on the Least squares estimate and the Modified Levenberg-Marquardt algorithm. The analytical derivation method is used for computation of the Jacobian matrix. In the developed algorithm, S_1 and S_2 represent the antecedent (non-linear) and consequent (linear) parameters respectively.

Let n represent the number of inputs,
Let p represent the number of membership functions for each input
Let l represent the layers of the proposed method, where $l = \{1, 2, 3, 4, 5\}$
Let the output of node i of layers l be O_i^j.

4.6 Forward Pass

Least squares estimate (LSE) was used at the very beginning to get the initial values of the consequent parameters $S_2 = \{p_i, q_i, r_i\}$.

Layer 1: Calculate the Membership Functions values for inputs. The Gaussian activation function was used as fuzzification nodes. The output of this layer is O_i^1, where $i = \{1, 2, ..., \text{p.n}\}$. O_i^1 is a membership function that satisfies the degree to which the given input satisfies the fuzzy sets A_i for $i = \{1, 2, ..., p\}$. The fuzzy sets are represented as a membership functions. The functions are expressed as $\mu_{A_i}(x_t; \{c, \sigma\}) for = \{1, 2, ..., n\}$,

where input n features are grid partitioned into p membership functions.

For simplicity, the output of two fuzzy membership grade of inputs, which are given by:

$$O_i^1 = \mu_{Ai}(x) = e^{-\frac{1}{2}\left(\frac{-c_i}{\sigma_i}\right)}, i = 1, 2 \qquad (4.2)$$

$$O_i^1 = \mu_{Bi}(y) = e^{-\frac{1}{2}\left(\frac{y-m_i}{\beta_i}\right)}, i = 1, 2 \qquad (4.3)$$

where x and y are the inputs to node i, $[c_i, \sigma_i, m_i, \beta_i]$ is a parameter set, represents the membership function's centres and widths of $\mu_{A_i}(x)$ and $\mu_{\beta_i}(x)$ respectively. The membership functions representing the antecedent parameters of the proposed method are described as $S_1 = \{a_i, b_i, c_i, ...\}$.

Let A_n represent the number of antecedent parameters for each membership function $\mu_{A_i}(x)$. The total number of antecedent parameters are then equivalent to:

$$Total\ number\ of\ antecedent\ parameters\ =\ A_p * (p * n) \qquad (4.4)$$

Layer 2: Calculate the rule firing strengths. Each node in this layer corresponds to a single Takagi-Sugeno type fuzzy rule. The output of this layer is O_i^2 where node $i = \{1, 2, ..., p^n\}$. The conjunction of rules antecedent is evaluated by either of the operator AND (minimum of incoming signals) or an OR (maximum of incoming signals).

Let R represent the rule choice of second layer nodes

$$R = \{\min\ [AND]\}\ or\ R\ =\ \{\max\ [OR]\} \qquad (4.5)$$

For simplicity, the output O_i^2 for two fuzzy IF-THEN rules is given by:

$$O_i^2 = w_i = rule\{A_i\} = \mu_{Ai}(x) * \mu_{Bi}(y) = e^{-\frac{1}{2}\left(\frac{-c_i}{\sigma_i}\right)} * e^{-\frac{1}{2}\left(\frac{y-m_i}{\beta_i}\right)}, i = 1, 2 \qquad (4.6)$$

where the value w_i represents the firing strength or weights from the rule node.

Layer 3: Determine the normalized firing strengths. The ratio of the firing strength of a given rule to the sum of firing strengths of all rules is called the normalized firing strength.

Let N represent the normalization of the node in layer 3. The output of this layer is O_i^3, where node $i = \{1, 2, ..., p^n\}$.

Let $\overline{w_i}$ represent the normalized weight of each rule

$$O_i^3\ =\ \overline{w_i}\ =\ \frac{w_i}{w_1 + w_2}\ =\ \frac{\mu_{Ai}(x) * \mu_{Bi}(y)}{\mu_{Ai}(x) + \mu_{Bi}(y)},\ i = 1, 2 \qquad (4.7)$$

Normalizing guarantees stable convergence of weights and biases as it avoids the time-consuming process of defuzzification.

Layer 4: This is a defuzzification layer. It calculates the rules outputs for rule consequent layer. Each node in this layer is an adaptive node. The output in this layer is simply the product of the normalized firing strength and a first order polynomial (for a first order TSK model). Thus, the output of this layer O_i^4, where node $i = \{1, 2, ..., p^n\}$.

Let $S_2 = [p_i, q_i, r_i]$ be the consequent parameters, which can be identified using least square estimation. A linear function f_i is expressed as a multiplication of the inputs with the corresponding consequent parameters. The output O_i^4 is the product of normalized firing strength $\overline{w_i}$ of layer 3 with the linear function f_i given by

$$O_i^4 = \overline{w_i} f_i = w_i\left(p_i x + q_i y + r_i\right),\ i = 1, 2, ...p^n \qquad (4.8)$$

$$The\ total\ number\ of\ consequent\ parameters\ =\ (n + 1) * p^n \qquad (4.9)$$

Layer 5: This is the summation layer. It is designed to calculate the sum of the output of all incoming signal, that's to computes the overall output as the summation of all incoming signals. The output of this layer is O_i^5, where node $i = \{1\}$. Since there is only one output, the ANFIS is a binary classifier. The output is the aggregation of all defuzzified outputs O_i^4 from layer 4, and thus it follows the weighted average.

$$O_i^5 = \sum_i \overline{w_i} f_i = \frac{\sum_i w_i f_i}{\sum_i w_i}, \quad i = 1, 2, ..., p^n \tag{4.10}$$

The least squares estimator is used to minimize the squared error $||\mathbf{AX} - \mathbf{B}||^2$, where $\mathbf{A} =$ Output produced by O_i^3, $\mathbf{y} =$ Target output and $\mathbf{X} =$ Unknown consequent values related to the set of consequent parameters p_i, q_i & r_i, which can be obtained using pseudo-inverse of $\mathbf{X}$.

Following Equation (4.10), we can develop an expression involving the normalized weights $\overline{w_i}$, Equation (4.7) multiplied by the inputs x_t, (layer 1), gives:

$$O_i^{5*} = \Sigma_i[(\overline{w_i}x)p_i + (\overline{w_i}y)q_i + \overline{w_i}r_i] \text{ for } i = \{1, 2, ..., p^n\} \tag{4.11}$$

After the consequent parameters S_2 are identified, the network output can be computed and the error measure E_k represents an objective function for kth of the training data can be obtained as:

$$\mathbf{E}_k = (\mathbf{t}_k - \mathbf{a}_k)^2 \tag{4.12}$$

where $\mathbf{t}_k$ and $\mathbf{a}_k$ represent the target output vector and actual output vector, and N is the number of total points. The overall error measure E of the training data set can be computed using performance measure, mean square error (MSE) defined as:

$$MSE = \frac{1}{N} \sum_{i=1}^{N} \mathbf{E}_k \tag{4.13}$$

4.7 Backward Pass

In the backward pass, error signals are propagated and antecedent parameters $S_1 = \{\sigma_i, c_i\}$ are to be updated by Modified Levenberg-Marquardt algorithm. The performance index to be optimized is defined by [10] and presented as:

$$F(\mathbf{w}) = \frac{1}{2}\mathbf{e}^T\mathbf{e} \tag{4.14}$$

$F(\mathbf{w})$ is the total error function, $\mathbf{w} = [w_1, w_2, ..., w_k]$ comprising of all weights of the network, $\mathbf{e}$ is the error vector comprising the error of all the training samples.

The parameters of unique membership functions of current fuzzy inference system (FIS) are to be obtained, which is a novel approach that allows program to run faster and is defined as:

$$\mathbf{v} = I(\mathbf{R}_{ij}) \tag{4.15}$$

where $\mathbf{v}$ is the rules index vector that keeps track of the unique membership functions MFs, I is the index table of the unique MF used in the rules, $\mathbf{R}_{ij}$ is a matrix of size number of rule by number of input, that identifies the membership functions for the ith rule and jth input.

4.7.1 Rules' Index Vector

The index Membership function is the index vector that keeps track of the unique Membership Functions (MFs). This function determines the unique MFs in the ANFIS structure and indexes them row-wise for an efficient classification system and to speed up convergence rates instead of using column-wise for conventional methods. Therefore, the final index vector collects the indices found in MF "row-wise" according to rules. For example, in Cleveland heart disease input variables with [3, 4, 3, 4, 2, 3, 3] the number of the membership function for each input respectively. The fourth input (cholesterol) has four membership functions defined as low, medium, high and very high. Now, consider a rule list:

$$\mathbf{R} = \begin{bmatrix} 1 & 1 & 1 & 1 & 1 & 1 & 1 \\ 2 & 2 & 2 & 2 & 2 & 2 & 2 \end{bmatrix} \tag{4.16}$$

$$\mathbf{ixMF} = \begin{bmatrix} 1 & 2 & 3 & 4 & 5 & 6 & 7 \\ 8 & 9 & 10 & 11 & 12 & 13 & 14 \end{bmatrix} \tag{4.17}$$

Look at the 4th column of the rule list, one can see that only the 1st (low) and 2nd (medium) MF is used. Unfortunately, MATLAB's fis structure numbers all MFs column-wise vector starting from the first input.

$$\mathbf{MF} = \begin{bmatrix} 1 & 4 & 8 & 11 & 15 & 17 & 20 \\ 2 & 5 & 9 & 12 & 16 & 18 & 21 \\ 3 & 6 & 10 & 13 & & 19 & 22 \\ & 7 & & 14 & & & \end{bmatrix} \tag{4.18}$$

where $\mathbf{R}$ is the rule list matrix [11].

As can be seen in the rule list the low-cholesterol **MF** is in the first row, fourth column. However, in the fis structure it has index of 11. This is what must keep track of. Thus, the final index vector collects the indices found in *MF* "row-wise" according to **R**.

first-row:
$R(1,1) = 1$, thus ix (1) = $MF(1,1) = 1$
$R(1,2) = 1$, thus ix (2) = $MF(1,2) = 4$
$R(1,3) = 1$, thus ix (3) = $MF(1,3) = 8$
$R(4,4) = 1$, thus ix(4.) = $MF(4,4) = 11$
$R(1,5) = 1$, thus ix (5) = $MF(1,5) = 15$
$R(1,6) = 1$, thus ix (6) = $MF(1,6) = 17$
$R(1,7) = 1$, thus ix (7) = $MF(1,7) = 20$

second-row:
$R(2,1) = 2$, thus ix (8) = $MF(2,1) = 2$
$R(2,2) = 2$, thus ix (9) = $MF(2,2) = 5$
$R(2,3) = 2$, thus ix (10) = $MF(2,3) = 9$
$R(2,4) = 2$, thus ix (11) = $MF(2,4) = 12$
$R(2,5) = 2$, thus ix (12) = $MF(2,5) = 16$
$R(2,6) = 2$, thus ix (13) = $MF(2,6) = 18$
$R(2,7) = 2$, thus ix (14) = $MF(2,7) = 21$

Finally, the rule index vector is:

$$\mathbf{v} = [1, \ 4, \ 8, \ 11, \ 15, \ 17, \ 20, \ 2, \ 5, \ 9, \ 12, \ 16, \ 18, \ 21] \qquad (4.19)$$

Again, let's consider a general case, using the same MFs for inputs 3 and 5. One can check this by the indexMF function:

$$\mathbf{R} = \begin{bmatrix} 1 & 1 & 1 & 4 & 2 & 1 & 1 \\ 3 & 2 & 1 & 2 & 2 & 3 & 3 \end{bmatrix} \qquad (4.20)$$

$$\mathbf{ixMF} = \begin{bmatrix} 1 & 2 & 3 & 4 & 5 & 6 & 7 \\ 8 & 9 & 3 & 10 & 5 & 11 & 12 \end{bmatrix} \qquad (4.21)$$

$$\mathbf{MF} = \begin{bmatrix} 1 & 4 & 8 & 11 & 15 & 17 & 20 \\ 2 & 5 & 9 & 12 & 16 & 18 & 21 \\ 3 & 6 & 10 & 13 & & 19 & 22 \\ & 7 & & 14 & & & \end{bmatrix} \qquad (4.22)$$

first-row:

$R(1,1) = 1$, thus ix (1) $= MF(1,1) = 1$
$R(1,2) = 1$, thus ix (2) $= MF(1,2) = 4$
$R(1,3) = 1$, thus ix (3) $= MF(1,3) = 8$
$R(1,4) = 4$, thus ix (4) $= MF(4,4) = 14$
$R(1,5) = 2$, thus ix (5) $= MF(2,5) = 16$
$R(1,6) = 1$, thus ix (6) $= MF(1,6) = 17$
$R(1,7) = 1$, thus ix (7) $= MF(1,7) = 20$

second-row:

$R(2,1) = 3$, thus ix (8) $= MF(3,1) = 3$
$R(2,2) = 2$, thus ix (9) $= MF(2,2) = 5$
$R(2,3) = $ **1, this is not unique, since it already used in rule 1, so SKIP**
$R(2,4) = 2$, thus ix (10) $= MF(2,4) = 12$
$R(2,5) = $ **2, this is not unique, since it already used in rule 1, so SKIP**
$R(2,6) = 3$, thus ix (11) $= MF(3,6) = 19$
$R(2,7) = 3$, thus ix (12) $= MF(3,7) = 22$

Finally, the index vector is (length of 12) unique MFs and NOT 14 in the other case.

$$\mathbf{v} = \begin{bmatrix} 1, & 4, & 8, & 14, & 16, & 17, & 20, & 3, & 5, & 12, & 19, & 22 \end{bmatrix} \tag{4.23}$$

The Jacobian matrix is built-up column–wise, which contains first order partial derivatives of network error using analytical derivation and chain rule method is given by:

$$\mathbf{J_k} = \frac{\partial f_i}{\partial \rho_j} = \begin{bmatrix} \partial y_1/\partial \sigma_1 & \partial y_1/\partial \beta_1 & \cdots & \partial y_1/\partial \sigma_{N_f} & \partial y_1/\partial \beta_{N_f} \\ \partial y_2/\partial \sigma_1 & \partial y_2/\partial \beta_1 & \cdots & \partial y_2/\partial \sigma_{N_f} & \partial y_2/\partial \beta_{N_f} \\ \vdots & \vdots & \ddots & \vdots & \vdots \\ \partial y_{N_t}/\partial \sigma_1 & \partial y_{N_t}/\partial \beta_1 & \cdots & \partial y_{N_t}/\partial \sigma_{N_f} & \partial y_2/\partial \beta_{N_f} \end{bmatrix}$$

$$\tag{4.24}$$

A novelty approach was introduced to transform Jacobian into sparse Jacobian matrix to speed things up. With sparse matrix storage, it is in general practical to store the rows of $\mathbf{J}_k$ in a compressed form, i.e. without zero entries [12].

$$\mathbf{J}_k = Sparse\left(\frac{\partial f_i}{\partial \rho_j}\right) \tag{4.25}$$

Using the chain rule, the gradient and Hessian of f(**x**) can be expressed in terms of the Jacobian,

$$F'(\mathbf{X}) = \mathbf{g} = \mathbf{J}_k^T \mathbf{e} \qquad (4.26)$$

Approximate Hessian matrix which contains second order partial derivative of network error using the cross product of Jacobian can be obtained as:

$$F''(\mathbf{X}) = \mathbf{H} \approx \mathbf{J}_k^T \mathbf{J}_k \qquad (4.27)$$

This approximation is used for both Gauss-Newton and Levenberg-Marquardt methods. Hence, explicit computation of second order derivatives can be avoided. Hessian matrix is not invertible as described in Gauss-Newton method. To ensure that equation is invertible, another approximation is introduced to the Hessian matrix and can be updated as:

$$\mathbf{H}^* = \mathbf{J}_k^T \mathbf{J}_k + \eta \mathbf{I} \qquad (4.28)$$

where η is called combination coefficient or learning parameter, $\mathbf{I}$ is the sparse identity matrix.

With the Modified Levenberg-Marquardt method (MLM), the increment of the parameter in training will be obtained as:

$$\Delta \mathbf{X}_k = \left(\mathbf{J}_k^T \mathbf{J}_k + \eta \mathbf{I}\right)^{-1} \mathbf{J}_k^T \mathbf{e} \qquad (4.29)$$

4.8 Simulation Results

Based on the five datasets obtained from the University of California Irvine (UCI) machine learning repository [13], throughout the experiments the ten-fold cross-validation method was used instead of the hold-out validation method as applied in the previous method [14]. The results with five data sets for robustness were compared based on the machine learning process in terms of accuracy, sensitivity and specificity [15].

The MLM algorithm in most cases is more effective and achieves a lower MSE and higher mapping precision. This is presented in Tables 4.1–4.5. The hyphen (—) means "Not Applicable" (NA), it indicates that there is no such type of result in the respective existing classifiers as presented in Tables 4.1–4.5.

The proposed classifier yields better results with faster convergence speed than other classifiers as presented in Table 4.1. The accuracy, sensitivity and specificity of the proposed classifier for Hepatitis datasets were obtained as 95.18%, 95.04% and 97.00% respectively. The MSE was found to be 0.0744 with an elapsed time of 12.63 seconds.

Table 4.1 Comparison of test accuracy results with some related existing models for Hepatitis data set

Methodology Adopted	Accuracy (%)	Sensitivity (%)	Specificity (%)	MSE	Elapsed Time (sec)
Proposed method	95.18	95.04	97.00	0.0744	12.63
Conventional method	93.33	93.20	94.26	0.0818	27.46
ABC-SVM [16]	94.92	97.13	88.33	—	—
CBR-PSO [17]	94.58	—	—	—	—
ABC-Boosting [18]	83.44	—	—	—	—

The accuracy, sensitivity and specificity of the proposed classifier were obtained as 96.34%, 98.29% and 89.61% respectively. The MSE was obtained as 0.0291 with an elapsed time of 2.99 seconds for SPECT Heart data set as presented in Table 4.2.

The proposed model yields better results than other models with much faster convergence speed as presented in Table 4.3 for Cleveland heart dataset. The accuracy, sensitivity and specificity of the proposed classifier were obtained as 79.71%, 67.06% and 80.38% respectively. The MSE was obtained as 0.1308 with an elapsed time of 0.64 seconds.

The proposed classifier yields better results with faster convergence speed than other classifiers as presented in Table 4.4. The accuracy, sensitivity and specificity of the proposed classifier for Diabetic Retinopathy Debrecen data set were obtained as 76.76%, 69.31% and 85.08% respectively. The MSE was found to be 0.1908 with an elapsed time of 4.83 seconds.

The proposed model yields better results than other models with much faster convergence speed as presented in Table 4.5 for BUPA Liver Disorders dataset. The accuracy, sensitivity and specificity of the proposed classifier were obtained as 72.17%, 78.95% and 68.97% respectively. The MSE was obtained as 0.4497 with an elapsed time of 9.83 seconds.

Table 4.2 Comparison of test accuracy results with some related existing models for SPECT Heart data set

Methodology Adopted	Accuracy (%)	Sensitivity (%)	Specificity (%)	MSE	Elapsed Time (sec)
Proposed method	96.34	98.29	89.61	0.0291	2.99
Conventional method	93.46	90.44	88.03	0.0293	9.06
RS [19]	93 ± 3.8	95.00	85.00	—	—
SMFFNN [20]	92.00	—	—	—	—
SBPN [21]	87.00	—	—	—	—
BPN+PCA [22]	73.30	—	—	—	—

Table 4.3 Comparison of test accuracy results with some related existing models for Cleveland Heart data set

Methodology Adopted	Accuracy (%)	Sensitivity (%)	Specificity (%)	MSE	Elapsed Time (sec)
Proposed method	79.71	67.06	80.38	0.1308	0.64
Conventional method	75.56	71.05	78.85	0.4003	1.95
ANFIS [23]	75.93	—	—	—	—
ANN [24]	65.00	—	—	—	56.50
ANN [25]	76.00	—	—	—	—

Table 4.4 Diabetic Retinopathy Debrecen data set

Methodology Adopted	Accuracy (%)	Sensitivity (%)	Specificity (%)	MSE	Elapsed Time (sec)
Proposed method	76.76	69.31	85.08	0.1908	4.83
Conventional method	75.98	68.32	84.53	0.1934	7.06
PCA+MGA [26]	66.00	—	—	—	—
ANN+ABC [27]	72.53	—	—	—	—

Table 4.5 BUPA Liver Disorders Data set

Methodology Adopted	Accuracy (%)	Sensitivity (%)	Specificity (%)	MSE	Elapsed Time (sec)
Proposed method	72.17	78.95	68.97	0.4497	9.83
Conventional method	70.43	77.63	66.41	0.4515	18.06
PSO+1-NN [28]	68.99	—	—	—	—
ABC+SVM [29]	74.81	—	—	—	—

4.9 Graphs of Performance Error Vs No. of Iterations

In each figure, there are two curves which show the proposed learning algorithm in blue and the conventional method in red. Both are trained simultaneously with the same number of iterations. The proposed plan can produce a more significant improvement in the error rate in some cases. The high number of iterations signifies that the network proceeds to approximate convergence. This indicates that some sense of training set adequately attains small training errors quickly and efficiently enough to generalise unseen data (test data) or instances.

The aim is to minimise both the false negative (which may let a disease go untreated and get worse) without inflating false positives too much (which may be very costly for patients) and being able to attain lowest measures. This allows the algorithm to reach a stable solution on each of the errors (MSE) in the least amount of time. Below are the figures for each of the datasets.

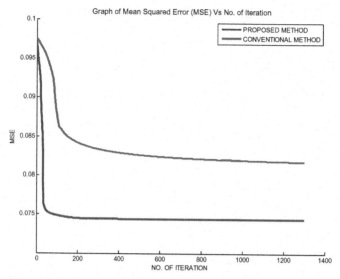

Figure 4.1 Graph of MSE vs. no. of iteration for hepatitis data set.

In Figure 4.1, it appears that the proposed method outperforms the conventional method as error decreases per training examples of the network and continues to drop. The error starts stabilising after 200 iterations for the proposed method and after 350 iterations for the conventional method. This is because the traditional approach has weaker convergence rates as the number of iterations is very high. The training process does not overfit the training data, and the proposed method has gained and produced the estimation of generalization in a final error achieved of 0.0744 at 12.63 seconds against 0.0818 at 27.46 seconds of the conventional method, both after 1300 iterations.

In Figure 4.2, it appears that the proposed method outperforms the conventional method as error decreases per training examples of the network and continues to drop. The error starts stabilizing after 50 iterations for the proposed method and after 125 iterations for the conventional method. The training process does not overfit the training data, and the proposed method has gained and produced the estimation of generalization in a final error achieved of 0.0291 at 2.99 seconds against 0.0293 at 9.06 seconds of the conventional method, both after 750 iterations.

In Figure 4.3, it appears that the proposed method outperforms the conventional method as error decreases per training examples of the network and continues to drop. The error starts stabilising after 150 and 230 iterations

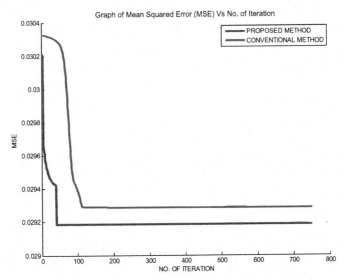

Figure 4.2   Graph of MSE vs. no. of iteration for SPECT-heart data set.

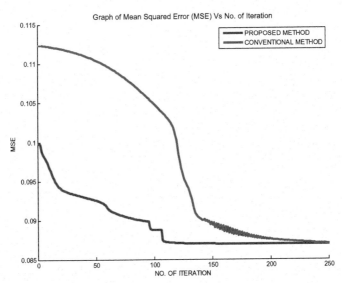

Figure 4.3   Graph of MSE vs. no. of iteration for cleveland heart data set.

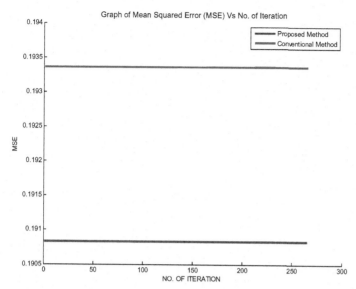

Figure 4.4   Graph of MSE vs. no. of iteration for diabetic retinopathy debrecen data set.

for the proposed and conventional methods respectively. The training process does not overfit the training data, and the proposed method has gained and produced the estimation of generalisation in a final error achieved of 0.1308 at 0.64 seconds against 0.4003 at 1.95 seconds of the conventional method, both after 250 iterations.

In Figure 4.4, it appears that errors of the proposed and conventional methods start stabilizing at the initial stages. The training process does not overfit the training data, and the proposed plan has gained and produced the estimation of generalization in a final error achieved of 0.1908 at 4.83 seconds against 0.1934 at 7.06 seconds of the conventional method, both after 265 iterations.

In Figure 4.5, it appears that the proposed method outperforms the conventional method as error decreases per training examples of the network and continues to drop. The error starts stabilising after 200 iterations for the proposed and traditional methods. The training process does not overfit the training data, and the proposed method has gained and produced the estimation of generalization in a final error achieved of 0.4497 at 9.83 seconds against 0.4515 at 18.06 seconds of the conventional method, both after 1200 iterations.

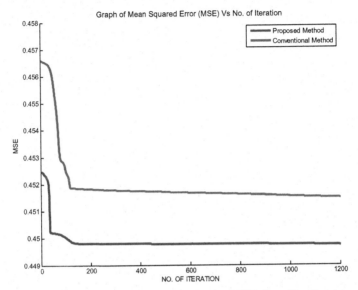

Figure 4.5   Graph of MSE vs. no. of iteration for BUPA liver disorders data set.

In conclusion, the proposed algorithm based on novelty techniques utilizes the application of the adaptive neuro-fuzzy inference system aimed at developing an effective, intelligence system that can be used by physicians to accelerate the diagnosis process. The proposed algorithm has achieved above performance error, reduced memory consumption or storage space has faster convergence speed and improved execution time.

References

[1] Sadek, "Artificial intelligence in transportation: information for application," Transportation Research Board Circular (E-C113), TRB, National Research Council, Washington, DC Available online at: http://onlinepubs.trb.org/onlinepubs/circulars/ec113. pdf (4.007).

[2] Buragohain, "Adaptive Network based Fuzzy Inference System (ANFIS) as a Tool for System Identification with Special Emphasis on Training Data Minimization," in *Book Adaptive Network based Fuzzy Inference System (ANFIS) as a Tool for System Identification with Special Emphasis on Training Data Minimization,* eds. Editor (City: Indian Institute of Technology Guwahati, 2009).

[3] Sharma and Kaur, "Classification in pattern recognition: A review," International Journal of Advanced Research in Computer Science and Software Engineering 3, no. 4 (2013).

[4] McCulloch and Pitts, "A logical calculus of the ideas immanent in nervous activity," The bulletin of mathematical biophysics 5, no. 4 (1943).

[5] McCulloch and Pitts, "A logical calculus of the ideas immanent in nervous activity," Bulletin of mathematical biology 52, no. 1–2 (1990).

[6] Galushkin, *Neural networks theory*, eds. Editor (Springer Science & Business Media, 2007.

[7] Basu, Bhattacharyya and Kim, "Use of artificial neural network in pattern recognition," International journal of software engineering and its applications 4, no. 2 (2010).

[8] Song, Zhu, Zheng, Tao and Wang, "Classifier Selection for Locomotion Mode Recognition Using Wearable Capacitive Sensing Systems," in *Robot Intelligence Technology and Applications 2*, (Springer, 2014).

[9] Sagir and Saratha, "A Novel Adaptive Neuro Fuzzy Inference System Based Classification Models for Heart Disease Prediction," Pertanika Journal of Science and Technology 25, no. 1 (2017).

[10] Madsen, Nielsen and Tingleff, "Methods for non-linear least squares problems [Lecture notes]. Retrieved 12 January, 2016," in Book Methods for non-linear least squares problems [Lecture notes]. Retrieved 12 January, 2016, eds. Editor (City, 2004).

[11] Jamsandekar and Mudholkar, "Fuzzy Classification System by Self Generated Membership Function Using Clustering Technique," BVICAM's International Journal of Information Technology 6, no. 1 (2014).

[12] Bunch and Rose, *Sparse matrix computations*, eds. Editor (Academic Press, 2014).

[13] Bache and Lichman, "UCI machine learning repository," in *Book UCI machine learning repository*, eds. Editor (City, 2013).

[14] Sagir, "An Integrated Fuzzy Model for Pattern Recognition," in *Book An Integrated Fuzzy Model for Pattern Recognition*, eds. Editor (City: Universiti Sains Malaysia, 2017).

[15] Saed, "An Introduction to Data Mining," in *Book An Introduction to Data Mining*, eds. Editor (City, 2015).

[16] Uzer, Yilmaz and Inan, "Feature selection method based on artificial bee colony algorithm and support vector machines for medical datasets classification," The Scientific World Journal 2013 (2013).

[17] Neshat, Sargolzaei, Nadjaran Toosi and Masoumi, "Hepatitis disease diagnosis using hybrid case based reasoning and particle swarm optimization," ISRN Artificial Intelligence 2012 (2012).

[18] Palanisamy and Kanmani, "Artificial bee colony approach for optimizing feature selection," International Journal of Computer Science 9, no. 3 (2012).

[19] Ibid.

[20] Asadi R, "New supervised multi-layer feed forward neural network (SMFFNN) model to accelerate classification with high accuracy based on SPECT and SPECTF heart disease data sets," European Journal of Scientific Research 33, no. 1 (2009).

[21] Ibid.

[22] Ibid.

[23] Uzer, Yilmaz and Inan, "Feature selection method based on artificial bee colony algorithm and support vector machines for medical datasets classification."

[24] El-Rafaie S, "On the use of SPECT imaging datasets for automated classification of ventricular heart disease," in *Book On the use of SPECT imaging datasets for automated classification of ventricular heart disease*, eds. Editor (City: Informatics and Systems (INFOS), 2012).

[25] Rani and Deepa, "Design of optimal fuzzy classifier system using particle swarm optimization," in *Book Design of optimal fuzzy classifier system using particle swarm optimization*, eds. Editor (City: IEEE, 2010).

[26] Sivasankar, Nair and Judy, "Feature Reduction in Clinical Data Classification using Augmented Genetic Algorithm," International Journal of Electrical and Computer Engineering 5, no. 6 (2015).

[27] Beloufa and Chikh, "Design of fuzzy classifier for diabetes disease using Modified Artificial Bee Colony algorithm," Computer methods and programs in biomedicine 112, no. 1 (2013).

[28] Chen, Su, Chen and Wang, "Particle swarm optimization for feature selection with application in obstructive sleep apnea diagnosis," Neural Computing and Applications 21, no. 8 (2012).

[29] Uzer, Yilmaz and Inan, "Feature selection method based on artificial bee colony algorithm and support vector machines for medical datasets classification."

5

Fuzzy Inventory Model with Demand, Deterioration and Inflation: A Comparative Study Through NGTFN and CNTFN

Magfura Pervin and Sankar Kumar Roy[*]

Department of Applied Mathematics with Oceanology and Computer Programming, Vidyasagar University, Midnapore-721102, West Bengal, India
E-mail: mamoni2014@rediffmail.com; sankroy2006@gmail.com
*Corresponding Author

In this chapter, an inventory model is developed where demand and deterioration rate are considered in fuzzy environment, because, in real-life situations, demand and deterioration rate of an item are slightly disrupted from their original values. Moreover, due to the political and business instability and uncertainty in the value of money, the fuzzy inflation rate is also incorporated in this model. The crisp model is fuzzified by using Normalized General Triangular Fuzzy Number (NGTFN) and Cloudy Normalized Triangular Fuzzy Number (CNTFN). Also, the total cost is calculated for both the crisp model and corresponding fuzzified model. Then, the costs are compared extensively among the crisp model and consequently the triangular fuzzy model and cloudy fuzzy model. A numerical example is included to justify the usefulness of the proposed model. A sensitivity analysis of the optimal solutions for the parameters is also provided. The paper ends with conclusions and an outlook to possible future studies.

5.1 Introduction

In classical economic order quantity (EOQ) model, the demand rate was regarded as constant. Harris [11] was the first to introduce the concept of constant demand in a deterministic model. Later, Pervin et al. [18, 20, 22] extended the concept of constant demand by considering linear decreasing demand, stock-dependent demand, stock and price dependent demand, respectively. Roy et al. [24] presented the deterministic demand as probabilistic in nature. But, in a real-life situation, the price of an item and demand are highly related to each other. If the demand for an item becomes high, the item will be produced in a large number, and as a result, there is a possibility that the unit production cost of the item may be decreased. Therefore, prediction of the exact amount of demand is a difficult job and it may increase or decrease with time to time. Hence, in this chapter, demand is observed as a decision variable and fuzzy in nature. Zadeh [32] was the first to introduce the concept of fuzzy set theory. Later on, Bellman and Zadeh [3] used this for solving decision-making problems for industrial purpose. Thereafter, Kaufmann and Gupta [13], Mahata and Goswami [15], Ban and Coroianu [2], Wu et al. [28] have studied the field extensively. Kao and Hsu [12] explained a lot size model with fuzzy demand. But to explore more results about fuzzy demand, we introduce this chapter.

Deterioration is a natural phenomenon which is found on almost all items like vegetables, food items, fresh fish, fashion items, drugs, radioactive substances, chemicals, and electronic components, while keeping in store. Deterioration is defined as a process of becoming worsened day by day. It is also falling from a higher to a lower level in quality. The effect of constant deterioration in an inventory model was nicely described by Pervin et al. [19, 21]. Generally, it is seen that deterioration has a small deviation from its original value in nature. Based on this argument, Vahidian and Tareghian [26] constructed an inventory model in a fuzzy environment. De and Goswami [4] derived an EPQ model where the deterioration was represented by a fuzzy number. In this chapter, we have also examined the effect of deterioration by allowing fuzziness in nature.

Inflation and time value of money play a significant role in an inventory model for an optimal ordering policy. Inflation always influences the demand for a certain product. It is found that many companies are suffering from high inflation, which means a huge loss of time value of money. As a result, dependent countries calculate loss for purchasing oil or power. When the time value of money collapses, spending on luxury items or peripherals always increases the demand. In this regard, several researchers like Skouri and Papachristos [25], Moon and Lee [17], Pervin et al. [23], validated their research works to execute new results in this direction. Since uncertainty and inflation are correlated with each other, therefore fuzzy inflation adds a positive impact in an inventory model. De and Goswami [5] developed a model by using this effect with fuzzy deterioration rate and allowed a delay in payments and extended the model of Yang et al. [31]. More are to be explored and we develop our model by adding fuzzy demand and fuzzy deterioration with fuzzy inflation.

Mahata [16] scrutinized the learning effect for imperfect production in fuzzy random environments in a study. Then, Kazemi et al. [14] derived the learning and forgetting effects of fuzzy parameters in an inventory model. It was assumed that the model will fail if the learning rate lies between [0–50%], and therefore, for the defuzzification method, ranking fuzzy numbers were introduced by Yager [30]. Then, the method was adopted by Ezzati et al. [9], Deng [8] and Allahviranloo and Saneifard [1], etc. Thereafter, De and Mahata [7] extended the work under cloudy fuzzy indices. Therefore, the works were extended over L-R fuzzy numbers by Hajjari and Abbasbandy [10], Wang et al. [27], Xu et al. [29] and others. Moreover, De and Beg [6] invented new defuzzification methods. This method can be applied in Neutrosophic sets, disaster management, environmental risk analysis, and even crime research also. We have utilized this extension to convert triangular dense fuzzy number to triangular cloudy fuzzy number and have applied it in inventory model with real-life applications.

The rest of the chapter is organized as follows: A preliminary concept about fuzzy numbers and cloudy fuzzy numbers are provided in Section 5.2. In Section 5.3, some notations about proposed model are presented. A clear mathematical model with fuzzy model and its corresponding solution is discussed in Section 5.4. Section 5.5 contains a numerical example to illustrate the model with a real-life example. Section 5.6 gives a sensitivity analysis with illustrated figures. Section 5.7 presents concluding remarks; and proposes a new pathway of future investigations.

5.2 Preliminary Concept

Fuzzy Set [32]

A fuzzy set $\tilde{A}$ on the universal set X is a set of ordered pair defined by $\tilde{A} = \{(x, \mu_{\tilde{A}}(x)) : x \in X\}$ where $\mu_{\tilde{A}} : X \rightarrow [0, 1]$ is called the membership function. The α-cut of the fuzzy set $\tilde{A}$ is presented as $A_\alpha = \{x : \mu_{\tilde{A}}(x) = \alpha, \alpha \in [0, 1]\}$. A fuzzy number is a fuzzy set $\tilde{A}$ together with the membership function $\mu_{\tilde{A}} : X \rightarrow [0, 1]$ with the following properties:

i. $\tilde{A}$ is normal, therefore, there exists a $x \in \mathbb{R}$ such that $\mu_{\tilde{A}}(x) = 1$.

ii. $\tilde{A}$ is piecewise continuous.

iii. $\sup(\tilde{A}) = cl\{x \in \mathbb{R} : \mu_{\tilde{A}}(x) > 0\}$, where cl represents the closure of a set.

iv. $\tilde{A}$ is a convex fuzzy set.

Normalized General Triangular Fuzzy Number (NGTFN)

Let B be a NGTFN of the form $\tilde{B} = (B_1, B_2, B_3)$. Then its membership function can be defined by $\mu(\tilde{B}) =$

$$
\begin{cases}
0, & \text{if } B < B_1 \text{ and } B > B_2 \\
\frac{B-B_1}{B_2-B_1}, & \text{if } B_1 \leq B \leq B_2 \\
\frac{B_3-B}{B_3-B_2}, & \text{if } B_2 \leq B \leq B_3
\end{cases}
\tag{5.1}
$$

Now, the left and right α-cut of $\mu(\tilde{B})$ are given by $L(\alpha) = B_1 + (B_2 - B_1)\alpha$ and $R(\alpha) = B_3 - (B_3 - B_2)\alpha$.

Yager's [30] Ranking Index

If $L(\alpha)$ and $R(\alpha)$ are the left and right spreads of a fuzzy number $\tilde{B}$, then the defuzzication rule under Yager's [30] ranking index is given by

$$
I(\tilde{B}) = \frac{1}{2} \int_0^1 [L(\alpha) + R(\alpha)]\, d\alpha = \frac{1}{4}(B_1 + 2B_2 + B_3).
\tag{5.2}
$$

The measure of fuzziness (degree of fuzziness d_f) is given by $d_f = \frac{U_b - L_b}{2m_1}$, where L_b and U_b are the lower bound and upper bound of the fuzzy number respectively, and m_1 being their respective modes.

Cloudy Normalized Triangular Fuzzy Number (CNTFN) (Extension of De and Beg [6])

A fuzzy number of the form $\tilde{B} = <b_1, b_2, b_3>$ is said to be a cloudy triangular fuzzy number if after an infinite time, the set itself converges to a crisp singleton. This implies that, as time $t \rightarrow \infty$, both $b_1, b_3 \rightarrow b_2$.

Let us consider the fuzzy number

$$\tilde{B} = <b_2\left(1 - \frac{\rho}{1+t}\right), \; b_2, \; b_2\left(1 + \frac{\sigma}{1+t}\right)>, \qquad (5.3)$$

for $0 < \rho, \sigma < 1$.

Note that, $lim_{t\to\infty} b_2\left(1 - \frac{\rho}{1+t}\right) = b_2$ and $lim_{t\to\infty} b_2\left(1 + \frac{\sigma}{1+t}\right) = b_2$, so $\tilde{B} \to \{b_2\}$.

Then the membership function for $t \geq 0$ is represented as follows:

$$\mu(x,t) = \begin{cases} 0, \text{ if } x < b_2\left(1 - \frac{\rho}{1+t}\right) \text{ and } x > b_2\left(1 + \frac{\sigma}{1+t}\right) \\[2mm] \frac{x - b_2\left(1 - \frac{\rho}{1+t}\right)}{\frac{\rho b_2}{1+t}}, \text{ if } b_2\left(1 - \frac{\rho}{1+t}\right) \leq x \leq b_2 \\[2mm] \frac{b_2\left(1 + \frac{\sigma}{1+t}\right) - x}{\frac{\rho b_2}{1+t}}, \text{ if } b_2 \leq x \leq b_2\left(1 + \frac{\sigma}{1+t}\right) \end{cases} \qquad (5.4)$$

Extended De and Beg's [6] Ranking Index on CNTFN

Let us consider the left and right α-cuts of $\mu(x,t)$ from Equation (5.4) and denoted it as L(α, t) and R(α, t) respectively. Then, the defuzzification formula under time extension of Yager's [30] ranking index is given by

$$I(\tilde{B}) = \frac{1}{2T} \int_{\alpha=0}^{1} \int_{t=0}^{T} \left[L^{-1}(\alpha,\,t) + R^{-1}(\alpha,\,t) \right] d\alpha \; dt \qquad (5.5)$$

where α and t are independent variables.

Let $\tilde{B}$ be a CNTFN defined in Equation (5.3) and its membership function is defined in Equation (5.4). Therefore, the left and right α-cuts of $\mu(x,t)$ from Equation (5.4), we get

$$L^{-1}(\alpha,\,t) = b_2\left(1 - \frac{\rho}{1+t} + \frac{\rho\alpha}{1+t}\right) \text{ and } R^{-1}(\alpha,\,t) = b_2\left(1 + \frac{\sigma}{1+t} - \frac{\sigma\alpha}{1+t}\right).$$

Then, from Equation (5.5), we have

$$I(\tilde{B}) = \frac{b_2}{2T}\left[2T + \frac{\sigma - \rho}{2}\log(1 + T)\right] \qquad (5.6)$$

Equation (5.6) can be rewritten as $I(\tilde{B}) = b_2\left[1 + \frac{\sigma - \rho}{4}\frac{\log(1+T)}{T}\right]$.

Obviously, $lim_{T\to\infty} \frac{\log(1+T)}{T} = 0$ and hence $I(\tilde{B}) \to b_2$ as $T \to \infty$. The factor $\frac{\log(1+T)}{T}$ is known as cloud index (CI) and in general, time T is measured by days.

5.3 Notations

The following notations are used to formulate the mathematical model throughout the chapter:

T	Length of cycle time (a decision variable);
k	Finite inflation rate;
$\tilde{k}$	Fuzzy inflation rate;
D	Demand rate per unit time, is time dependent and expressed as $D = a + bt$, where $a,\ b \in \mathbb{R}$ and $b \neq 0$;
$\tilde{D}$	Fuzzy demand rate per unit time;
θ	Unit deterioration rate;
$\tilde{\theta}$	Unit fuzzy deterioration rate;
m	Finite replenishment rate;
α	Unit fraction of imperfectness of production;
$I_1(t)$	Inventory level with respect to time during production period;
$I_2(t)$	Inventory level with respect to time during the non-production period;
h_1	Unit holding cost for non-defective items per unit per time unit;
h_2	Unit holding cost for defective items per unit per time unit;
c	Unit purchasing cost per item;
q	Size of each shipment from supplier to the retailer;
Q	Retailer's order quantity per replenishment (a decision variable);
$A(t)$	Unit ordering cost per item which is expressed as $A(t) = A_0(1 + kt)$, where A_0 is the constant, ordering cost per item;
r_1	Unit inspection cost per item per time unit;
r_2	Unit transportation cost of a shipment from supplier to the retailer;
$Z(t)$	Total average inventory cost per unit time (a decision variable);

5.4 Mathematical Model

The inventory cycle starts with a zero stock level at supply rate m among which α units are imperfect due to faulty machines and deterioration.

The replenishment or supply continues up to time t_1. During the time period $[0,\ t_1]$, the inventory piles up due to the demand in the market and stops at time $t = t_1$. This accumulated inventory level at time t_1 gradually diminishes due to demand and deterioration during the period $[t_1,\ T]$ and ultimately falls to 0 at the end of the cycle period T. After the scheduling period $[0,\ T]$, the cycle repeat itself.

Now, the differential equations involving the instantaneous state of the inventory level in the interval $[0,\ T]$, together with their initial values, are given subsequently:

$$\frac{dI_1(t)}{dt} + \theta I_1(t) = m(1-\alpha) - D, \qquad 0 \le t \le t_1 \tag{5.7}$$

with the initial condition $I_1(0) = 0$.
  And

$$\frac{dI_2(t)}{dt} + \theta I_2(t) = -(a+bt), \qquad t_1 \le t \le T \tag{5.8}$$

with the boundary condition $I_2(T) = 0$.
  Now, the solution of Equation (5.7) using the initial condition becomes

$$I_1(t) = \left(\frac{m(1-\alpha)-a}{\theta} + \frac{b}{\theta^2}\right)(1-e^{-\theta t}) - \frac{bt}{\theta}, \qquad 0 \le t \le t_1.$$

Utilizing the boundary condition, the solution of Equation (5.8) becomes

$$I_2(t) = \left(\frac{a}{\theta} - \frac{b}{\theta^2} + \frac{bT}{\theta}\right)e^{-\theta t} - \left(\frac{a+bt}{\theta} - \frac{b}{\theta^2}\right), \qquad t_1 \le t \le T.$$

Here $Q = m(1-\alpha)DT$.
  Figure 5.1 represents the graphical form of the inventory model.

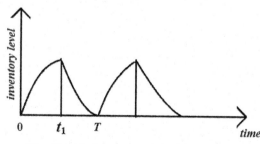

Figure 5.1 Graphical representation of the inventory model.

The elements which take part in calculating the cost function are listed below:

a. Annual ordering cost, OC, is $= qA_0(1 + kt)$;

b. Total holding cost for non-defective items, HCN, is computed as follows:

$$HCN = qh_1(1 - \alpha) \left[\int_0^{t_1} I_1(t)e^{-kt}dt + \int_{t_1}^T I_2(t)e^{-kt}dt \right]$$

$$= qh_1(1 - \alpha) \left[\left(\frac{m(1 - \alpha) - a}{\theta} + \frac{b}{\theta^2} \right) \left(\frac{e^{-(\theta+k)t_1} - 1}{\theta + k} - \frac{e^{-kt_1} - 1}{k} \right) \right.$$

$$+ \frac{b}{\theta} \left(\frac{t_1 e^{-kt_1}}{k} + \frac{e^{-kt_1} - 1}{k^2} \right) + \left(\frac{a + bT}{\theta} - \frac{b}{\theta^2} \right) \left(\frac{e^{-(\theta+k)t_1}}{\theta + k} - \frac{e^{-(\theta+k)T}}{\theta + k} \right)$$

$$\left. - \left(\frac{a}{\theta} - \frac{b}{\theta^2} \right) \left(\frac{e^{-kt_1}}{k} - \frac{e^{-kT}}{k} \right) + \frac{b}{\theta} \left(\frac{Te^{-kT} - t_1 e^{-kt_1}}{k} + \frac{e^{-kT} - e^{-kt_1}}{k^2} \right) \right].$$

c. Total holding cost for defective items, HCD, is calculated as follows:

$$HCD = qh_2\alpha \left[\int_0^{t_1} I_1(t)e^{-kt}dt + \int_{t_1}^T I_2(t)e^{-kt}dt \right]$$

$$= qh_2\alpha \left[\left(\frac{m(1 - \alpha) - a}{\theta} + \frac{b}{\theta^2} \right) \left(\frac{e^{-(\theta+k)t_1} - 1}{\theta + k} - \frac{e^{-kt_1} - 1}{k} \right) \right.$$

$$+ \frac{b}{\theta} \left(\frac{t_1 e^{-kt_1}}{k} + \frac{e^{-kt_1} - 1}{k^2} \right) + \left(\frac{a + bT}{\theta} - \frac{b}{\theta^2} \right) \left(\frac{e^{-(\theta+k)t_1}}{\theta + k} - \frac{e^{-(\theta+k)T}}{\theta + k} \right)$$

$$\left. - \left(\frac{a}{\theta} - \frac{b}{\theta^2} \right) \left(\frac{e^{-kt_1}}{k} - \frac{e^{-kT}}{k} \right) + \frac{b}{\theta} \left(\frac{Te^{-kT} - t_1 e^{-kt_1}}{k} + \frac{e^{-kT} - e^{-kt_1}}{k^2} \right) \right].$$

d. The deteriorating cost, DC, is represented by:

$$DC = qc\theta \left[\int_0^{t_1} (m(1 - \alpha) - (a + bt))e^{-kt}dt + \int_{t_1}^T I_2(t)e^{-kt}dt \right]$$

$$= qc\theta \left[\frac{m(1 - \alpha) - a}{k}(1 - e^{-kt_1}) + b \left(\frac{t_1 e^{-kt_1}}{k} + \frac{e^{-kt_1} - 1}{k^2} \right) \right.$$

$$+ \left(\frac{a + bT}{\theta} - \frac{b}{\theta^2} \right) \left(\frac{e^{-(\theta+k)t_1}}{\theta + k} - \frac{e^{-(\theta+k)T}}{\theta + k} \right) - \left(\frac{a}{\theta} - \frac{b}{\theta^2} \right) \left(\frac{e^{-kt_1}}{k} - \frac{e^{-kT}}{k} \right)$$

$$\left. + \frac{b}{\theta} \left(\frac{Te^{-kT} - t_1 e^{-kt_1}}{k} + \frac{e^{-kT} - e^{-kt_1}}{k^2} \right) \right].$$

e. For improving the production process and for maintaining the reputation of the business, the retailer follows an inspection policy to inspect all the defective items after receiving. Therefore, total inspection cost, IC, per replenishment cycle is given by:

$$IC = qr_1Q = q^2r_1m(1-\alpha).$$

f. Total transportation cost, TC, for all the items received is

$$TC = r_2Q = r_2qm(1-\alpha).$$

Therefore, the total cost of the system is expressed as: Total cost = OC+HCN+HCD+DC+IC+TC.

Hence, the total average cost is

$$Z(T,t_1) = \frac{qA_0}{T}(1+kt) + \frac{q}{T}(h_1(1-\alpha) + h_2\alpha)$$

$$\left[\left(\frac{m(1-\alpha)-a}{\theta} + \frac{b}{\theta^2}\right)\left(\frac{e^{-(\theta+k)t_1}-1}{\theta+k} - \frac{e^{-kt_1}-1}{k}\right)\right.$$

$$+\frac{b}{\theta}\left(\frac{t_1e^{-kt_1}}{k} + \frac{e^{-kt_1}-1}{k^2}\right) + \left(\frac{a+bT}{\theta} - \frac{b}{\theta^2}\right)\left(\frac{e^{-(\theta+k)t_1}}{\theta+k} - \frac{e^{-(\theta+k)T}}{\theta+k}\right)$$

$$-\left(\frac{a}{\theta} - \frac{b}{\theta^2}\right)\left(\frac{e^{-kt_1}}{k} - \frac{e^{-kT}}{k}\right) + \frac{b}{\theta}\left(\frac{Te^{-kT}-t_1e^{-kt_1}}{k} + \frac{e^{-kT}-e^{-kt_1}}{k^2}\right)\right]$$

$$+\frac{q}{T}c\theta\left[\frac{m(1-\alpha)-a}{k}(1-e^{-kt_1}) + b\left(\frac{t_1e^{-kt_1}}{k} + \frac{e^{-kt_1}-1}{k^2}\right)\right.$$

$$+\left(\frac{a+bT}{\theta} - \frac{b}{\theta^2}\right)\left(\frac{e^{-(\theta+k)t_1}}{\theta+k} - \frac{e^{-(\theta+k)T}}{\theta+k}\right) - \left(\frac{a}{\theta} - \frac{b}{\theta^2}\right)\left(\frac{e^{-kt_1}}{k} - \frac{e^{-kT}}{k}\right)$$

$$+\frac{b}{\theta}\left(\frac{Te^{-kT}-t_1e^{-kt_1}}{k} + \frac{e^{-kT}-e^{-kt_1}}{k^2}\right)\right] + \frac{q^2}{T}r_1m(1-\alpha) + \frac{q}{T}r_2m(1-\alpha).$$

Therefore, the problem can be represented as

$$\begin{cases} Minimi\ e\ Z(T,\ t_1) \\ subject\ to\ Q = m(1-\alpha)DT \end{cases} \tag{5.9}$$

5.5 Formulation of Fuzzy Mathematical Model

Let the demand, deterioration and inflation rate follows general fuzzy and cloudy fuzzy over the inventory run time and is represented as:

$$\tilde{D} = \begin{cases} <D_1,\ D_2,\ D_3> & \text{for } NGTFN \\ <D_2\left(1 - \frac{\rho}{1+T}\right),\ D_2,\ D_2\left(1 + \frac{\sigma}{1+T}\right)> & \text{for } CNTFN \end{cases}$$

$$\tilde{\theta} = \begin{cases} <\theta_1, \, \theta_2, \, \theta_3> & \text{for } NGTFN \\ <\theta_2\left(1 - \frac{\rho}{1+T}\right), \, \theta_2, \, \theta_2\left(1 + \frac{\sigma}{1+T}\right)> & \text{for } CNTFN \end{cases}$$

$$\tilde{k} = \begin{cases} <k_1, \, k_2, \, k_3> & \text{for } NGTFN \\ <k_2\left(1 - \frac{\rho}{1+T}\right), \, k_2, \, k_2\left(1 + \frac{\sigma}{1+T}\right)> & \text{for } CNTFN \end{cases}$$

where $0 < \rho, \, \sigma < 1$; $T > 0$ and $D_1 = a_1 + b_1 t$, $D_2 = a_2 + b_2 t$ and $D_3 = a_3 + b_3 t$.

Therefore, the corresponding fuzzy problem can be written as

$$\widetilde{Min} \, Z(T, t_1, \tilde{a}, \, \tilde{b}, \tilde{\theta}, \, \tilde{k}) = \frac{qA_0}{T}(1 + \tilde{k}t) + \frac{q}{T}(h_1(1 - \alpha) + h_2\alpha)$$

$$\left[\left(\frac{m(1-\alpha) - \tilde{a}}{\tilde{\theta}} + \frac{\tilde{b}}{\tilde{\theta}^2}\right)\left(\frac{e^{-(\tilde{\theta}+\tilde{k})t_1} - 1}{\tilde{\theta} + \tilde{k}} - \frac{e^{-\tilde{k}t_1} - 1}{\tilde{k}}\right)\right.$$

$$+ \frac{\tilde{b}}{\tilde{\theta}}\left(\frac{t_1 e^{-\tilde{k}t_1}}{\tilde{k}} + \frac{e^{-\tilde{k}t_1} - 1}{\tilde{k}^2}\right) + \left(\frac{\tilde{a} + \tilde{b}T}{\tilde{\theta}} - \frac{\tilde{b}}{\tilde{\theta}^2}\right)\left(\frac{e^{-(\tilde{\theta}+\tilde{k})t_1}}{\tilde{\theta} + \tilde{k}} - \frac{e^{-(\tilde{\theta}+\tilde{k})T}}{\tilde{\theta} + \tilde{k}}\right)$$

$$\left. - \left(\frac{\tilde{a}}{\tilde{\theta}} - \frac{\tilde{b}}{\tilde{\theta}^2}\right)\left(\frac{e^{-\tilde{k}t_1}}{\tilde{k}} - \frac{e^{-\tilde{k}T}}{\tilde{k}}\right) + \frac{\tilde{b}}{\tilde{\theta}}\left(\frac{Te^{-\tilde{k}T} - t_1 e^{-\tilde{k}t_1}}{\tilde{k}} + \frac{e^{-\tilde{k}T} - e^{-\tilde{k}t_1}}{\tilde{k}^2}\right)\right]$$

$$+ \frac{q}{T}c\tilde{\theta}\left[\frac{m(1-\alpha) - \tilde{a}}{\tilde{k}}(1 - e^{-\tilde{k}t_1}) + \tilde{b}\left(\frac{t_1 e^{-\tilde{k}t_1}}{\tilde{k}} + \frac{e^{-\tilde{k}t_1} - 1}{\tilde{k}^2}\right) + \left(\frac{\tilde{a} + \tilde{b}T}{\tilde{\theta}} - \frac{\tilde{b}}{\tilde{\theta}^2}\right)\right.$$

$$\left(\frac{e^{-(\tilde{\theta}+\tilde{k})t_1}}{\tilde{\theta} + \tilde{k}} - \frac{e^{-(\tilde{\theta}+\tilde{k})T}}{\tilde{\theta} + \tilde{k}}\right) - \left(\frac{\tilde{a}}{\tilde{\theta}} - \frac{\tilde{b}}{\tilde{\theta}^2}\right)\left(\frac{e^{-\tilde{k}t_1}}{\tilde{k}} - \frac{e^{-\tilde{k}T}}{\tilde{k}}\right)$$

$$\left. + \frac{\tilde{b}}{\tilde{\theta}}\left(\frac{Te^{-\tilde{k}T} - t_1 e^{-\tilde{k}t_1}}{\tilde{k}} + \frac{e^{-\tilde{k}T} - e^{-\tilde{k}t_1}}{\tilde{k}^2}\right)\right] + \frac{q^2}{T}r_1 m(1 - \alpha) + \frac{q}{T}r_2 m(1 - \alpha),$$

$$(5.10)$$

subject to

$$\tilde{Q} = m\,(1 - \alpha)\,\tilde{D}T \qquad\qquad (5.11)$$

Therefore, with the help of Equation (5.2), the membership function for the fuzzy objective function under NGTFN are given by

$$\mu_1(Z) = \begin{cases} 0, & \text{if } Z < Z_1 \text{ and } Z > Z_2 \\ \frac{Z - Z_1}{Z_2 - Z_1}, & \text{if } Z_1 \leq Z \leq Z_2 \\ \frac{Z_3 - Z}{Z_3 - Z_2}, & \text{if } Z_2 \leq Z \leq Z_3 \end{cases}$$

where

$$Z_1 = \frac{qA_0}{T}(1 + k_1 t) + \frac{q}{T}(h_1(1 - \alpha) + h_2\alpha)$$

$$\left[\left(\frac{m(1-\alpha) - a_1}{\theta_1} + \frac{b_1}{\theta_1^2}\right)\left(\frac{e^{-(\theta_1+k_1)t_1} - 1}{\theta_1 + k_1} - \frac{e^{-k_1 t_1} - 1}{k_1}\right)\right.$$

$$+\frac{b_1}{\theta_1}\left(\frac{t_1 e^{-k_1 t_1}}{k_1}+\frac{e^{-k_1 t_1}-1}{k_1^2}\right)+\left(\frac{a_1+b_1 T}{\theta_1}-\frac{b_1}{\theta_1^2}\right)$$

$$\left(\frac{e^{-(\theta_1+k_1)t_1}}{\theta_1+k_1}-\frac{e^{-(\theta_1+k_1)T}}{\theta_1+k_1}\right)-\left(\frac{a_1}{\theta_1}-\frac{b_1}{\theta_1^2}\right)\left(\frac{e^{-k_1 t_1}}{k_1}-\frac{e^{-k_1 T}}{k_1}\right)$$

$$+\frac{b_1}{\theta_1}\left(\frac{T e^{-k_1 T}-t_1 e^{-k_1 t_1}}{k_1}+\frac{e^{-k_1 T}-e^{-k_1 t_1}}{k_1^2}\right)\Bigg]$$

$$+\frac{q}{T}c\theta_1\Bigg[\frac{m(1-\alpha)-a_1}{k_1}(1-e^{-k_1 t_1})+b_1\left(\frac{t_1 e^{-k_1 t_1}}{k_1}+\frac{e^{-k_1 t_1}-1}{k_2^1}\right)$$

$$+\left(\frac{a_1+b_1 T}{\theta_1}-\frac{b_1}{\theta_1^2}\right)\left(\frac{e^{-(\theta_1+k_1)t_1}}{\theta_1+k_1}-\frac{e^{-(\theta_1+k_1)T}}{\theta_1+k_1}\right)-\left(\frac{a_1}{\theta_1}-\frac{b_1}{\theta_1^2}\right)$$

$$\left(\frac{e^{-k_1 t_1}}{k_1}-\frac{e^{-k_1 T}}{k_1}\right)+\frac{b_1}{\theta_1}\left(\frac{T e^{-k_1 T}-t_1 e^{-k_1 t_1}}{k_1}+\frac{e^{-k_1 T}-e^{-k_1 t_1}}{k_1^2}\right)\Bigg]$$

$$+\frac{q^2}{T}r_1 m(1-\alpha)+\frac{q}{T}r_2 m(1-\alpha),$$

$$Z_2=\frac{qA_0}{T}(1+k_2 t)+\frac{q}{T}(h_1(1-\alpha)+h_2\alpha)$$

$$\Bigg[\left(\frac{m(1-\alpha)-a_2}{\theta_2}+\frac{b_2}{\theta_2^2}\right)\left(\frac{e^{-(\theta_2+k_2)t_1}-1}{\theta_2+k_2}-\frac{e^{-k_2 t_1}-1}{k_2}\right)$$

$$+\frac{b_2}{\theta_2}\left(\frac{t_1 e^{-k_2 t_1}}{k_2}+\frac{e^{-k_2 t_1}-1}{k_2^2}\right)+\left(\frac{a_2+b_2 T}{\theta_2}-\frac{b_2}{\theta_2^2}\right)$$

$$\left(\frac{e^{-(\theta_2+k_2)t_1}}{\theta_2+k_2}-\frac{e^{-(\theta_2+k_2)T}}{\theta_2+k_2}\right)-\left(\frac{a_2}{\theta_2}-\frac{b_2}{\theta_2^2}\right)\left(\frac{e^{-k_2 t_1}}{k_2}-\frac{e^{-k_2 T}}{k_2}\right)$$

$$+\frac{b_2}{\theta_2}\left(\frac{T e^{-k_2 T}-t_1 e^{-k_2 t_1}}{k_2}+\frac{e^{-k_2 T}-e^{-k_2 t_1}}{k_2^2}\right)\Bigg]+\frac{q}{T}c\theta_2\Bigg[\frac{m(1-\alpha)-a_2}{k_2}$$

$$(1-e^{-k_2 t_1})+b_2\left(\frac{t_1 e^{-k_2 t_1}}{k_2}+\frac{e^{-k_2 t_1}-1}{k_2^2}\right)+\left(\frac{a_2+b_2 T}{\theta_2}-\frac{b_2}{\theta_2^2}\right)$$

$$\left(\frac{e^{-(\theta_2+k_2)t_1}}{\theta_2+k_2}-\frac{e^{-(\theta_2+k_2)T}}{\theta_2+k_2}\right)-\left(\frac{a_2}{\theta_2}-\frac{b_2}{\theta_2^2}\right)\left(\frac{e^{-k_2 t_1}}{k_2}-\frac{e^{-k_2 T}}{k_2}\right)$$

$$+\frac{b_2}{\theta_2}\left(\frac{T e^{-k_2 T}-t_1 e^{-k_2 t_1}}{k_2}+\frac{e^{-k_2 T}-e^{-k_2 t_1}}{k_2^2}\right)\Bigg]+\frac{q^2}{T}r_1 m(1-\alpha)+\frac{q}{T}r_2 m(1-\alpha)$$

and

$$Z_3=\frac{qA_0}{T}(1+k_3 t)+\frac{q}{T}(h_1(1-\alpha)+h_2\alpha)$$

$$\Bigg[\left(\frac{m(1-\alpha)-a_3}{\theta_3}+\frac{b_3}{\theta_3^2}\right)\left(\frac{e^{-(\theta_3+k_3)t_1}-1}{\theta_3+k_3}-\frac{e^{-k_3 t_1}-1}{k_3}\right)$$

$$+ \frac{b_3}{\theta_3}\left(\frac{t_1 e^{-k_3 t_1}}{k_3} + \frac{e^{-k_3 t_1} - 1}{k_3^2}\right) + \left(\frac{a_3 + b_3 T}{\theta_3} - \frac{b_3}{\theta_3^2}\right)$$

$$\left(\frac{e^{-(\theta_3 + k_3)t_1}}{\theta_3 + k_3} - \frac{e^{-(\theta_3 + k_3)T}}{\theta_3 + k_3}\right) - \left(\frac{a_3}{\theta_3} - \frac{b_3}{\theta_3^2}\right)\left(\frac{e^{-k_3 t_1}}{k_3} - \frac{e^{-k_3 T}}{k_3}\right)$$

$$+ \frac{b_3}{\theta_3}\left(\frac{T e^{-k_3 T} - t_1 e^{-k_3 t_1}}{k_3} + \frac{e^{-k_3 T} - e^{-k_3 t_1}}{k_3^2}\right)\right] + \frac{q}{T} c\theta_3 \left[\frac{m(1 - \alpha) - a_3}{k_3}\right.$$

$$(1 - e^{-k_3 t_1}) + b_3 \left(\frac{t_1 e^{-k_3 t_1}}{k_3} + \frac{e^{-k_3 t_1} - 1}{k_1^2}\right) + \left(\frac{a_3 + b_3 T}{\theta_3} - \frac{b_3}{\theta_3^2}\right)$$

$$\left(\frac{e^{-(\theta_3 + k_3)t_1}}{\theta_3 + k_3} - \frac{e^{-(\theta_3 + k_3)T}}{\theta_3 + k_3}\right) - \left(\frac{a_3}{\theta_3} - \frac{b_3}{\theta_3^2}\right)\left(\frac{e^{-k_3 t_1}}{k_3} - \frac{e^{-k_3 T}}{k_3}\right)$$

$$+ \frac{b_3}{\theta_3}\left(\frac{T e^{-k_3 T} - t_1 e^{-k_3 t_1}}{k_3} + \frac{e^{-k_3 T} - e^{-k_3 t_1}}{k_3^2}\right)\right] + \frac{q^2}{T} r_1 m(1 - \alpha) + \frac{q}{T} r_2 m(1 - \alpha).$$

$$\mu_2(Q) = \begin{cases} 0, & \text{if } Q < Q_1 \text{ and } Q > Q_2 \\ \frac{Q - Q_1}{Q_2 - Q_1}, & \text{if } Q_1 \le Q \le Q_2 \\ \frac{Q_3 - Q}{Q_3 - Q_2}, & \text{if } Q_2 \le Q \le Q_3 \end{cases}$$

where

$$\begin{cases} Q_1 = m(1 - \alpha)D_1 T \\ Q_2 = m(1 - \alpha)D_2 T \\ Q_3 = m(1 - \alpha)D_3 T \end{cases}$$

and the index values of the corresponding functions are given as

$$I(\tilde{Z}) = \frac{1}{4}(Z_1 + 2Z_2 + Z_3) = \frac{qA_0}{T}\left(1 + \frac{1}{4}(k_1 + 2k_2 + k_3)t\right) + \frac{q}{T}(h_1(1 - \alpha) + h_2\alpha)$$

$$\left(\frac{m(1 - \alpha) - \frac{1}{4}(a_1 + 2a_2 + a_3)}{\frac{1}{4}(\theta_1 + 2\theta_2 + \theta_3)} + \frac{\frac{1}{4}(b_1 + 2b_2 + b_3)}{\frac{1}{4}(\theta_1^2 + 2\theta_2^2 + \theta_3^2)}\right)$$

$$\left(\frac{e^{-\left(\frac{1}{4}(\theta_1 + 2\theta_2 + \theta_3) + \frac{1}{4}(k_1 + 2k_2 + k_3)\right)t_1} - 1}{\frac{1}{4}(\theta_1 + 2\theta_2 + \theta_3) + \frac{1}{4}(k_1 + 2k_2 + k_3)} - \frac{e^{-\frac{1}{4}(k_1 + 2k_2 + k_3)t_1} - 1}{\frac{1}{4}(k_1 + 2k_2 + k_3)}\right)$$

$$+ \frac{\frac{1}{4}(b_1 + 2b_2 + b_3)}{\frac{1}{4}(\theta_1 + 2\theta_2 + \theta_3)}\left(\frac{t_1 e^{-\frac{1}{4}(k_1 + 2k_2 + k_3)t_1}}{\frac{1}{4}(k_1 + 2k_2 + k_3)} + \frac{e^{-\frac{1}{4}(k_1 + 2k_2 + k_3)t_1} - 1}{\frac{1}{4}(k_1 + 2k_2 + k_3)^2}\right)$$

$$+ \left(\frac{\frac{1}{4}(a_1 + 2a_2 + a_3) + b_2 T}{\frac{1}{4}(\theta_1 + 2\theta_2 + \theta_3)} - \frac{\frac{1}{4}(b_1 + 2b_2 + b_3)}{\frac{1}{4}(\theta_1^2 + 2\theta_2^2 + \theta_3^2)}\right)$$

$$\left(\frac{e^{-\left(\frac{1}{4}(\theta_1 + 2\theta_2 + \theta_3) + \frac{1}{4}(k_1 + 2k_2 + k_3)\right)t_1}}{\frac{1}{4}(\theta_1 + 2\theta_2 + \theta_3) + \frac{1}{4}(k_1 + 2k_2 + k_3)} - \frac{e^{-\left(\frac{1}{4}(\theta_1 + 2\theta_2 + \theta_3) + \frac{1}{4}(k_1 + 2k_2 + k_3)\right)T}}{\frac{1}{4}(\theta_1 + 2\theta_2 + \theta_3) + \frac{1}{4}(k_1 + 2k_2 + k_3)}\right)$$

$$-\left(\frac{\frac{1}{4}(a_1+2a_2+a_3)}{\frac{1}{4}(\theta_1+2\theta_2+\theta_3)}-\frac{\frac{1}{4}(b_1+2b_2+b_3)}{\frac{1}{4}(\theta_1^2+2\theta_2^2+\theta_3^2)}\right)$$

$$\left(\frac{e^{-\frac{1}{4}(k_1+2k_2+k_3)t_1}}{\frac{1}{4}(k_1+2k_2+k_3)}-\frac{e^{-\frac{1}{4}(k_1+2k_2+k_3)T}}{\frac{1}{4}(k_1+2k_2+k_3)}\right)+\frac{\frac{1}{4}(b_1+2b_2+b_3)}{\frac{1}{4}(\theta_1+2\theta_2+\theta_3)}$$

$$\left(\frac{Te^{-\frac{1}{4}(k_1+2k_2+k_3)T}-t_1e^{-\frac{1}{4}(k_1+2k_2+k_3)t_1}}{\frac{1}{4}(k_1+2k_2+k_3)}\right.$$

$$\left.+\frac{e^{-\frac{1}{4}(k_1+2k_2+k_3)T}-e^{-\frac{1}{4}(k_1+2k_2+k_3)t_1}}{\frac{1}{4}(k_1^2+2k_2^2+k_3^2)}\right)+\frac{q}{T}c\frac{1}{4}(\theta_1+2\theta_2+\theta_3)$$

$$\left[\frac{m(1-\alpha)-\frac{1}{4}(a_1+2a_2+a_3)}{\frac{1}{4}(k_1+2k_2+k_3)}(1-e^{-\frac{1}{4}(k_1+2k_2+k_3)t_1})+\frac{1}{4}(b_1+2b_2+b_3)\right.$$

$$\left(\frac{t_1e^{-\frac{1}{4}(k_1+2k_2+k_3)t_1}}{\frac{1}{4}(k_1+2k_2+k_3)}+\frac{e^{-\frac{1}{4}(k_1+2k_2+k_3)t_1}-1}{\frac{1}{4}(k_1^2+2k_2^2+k_3^2)}\right)$$

$$+\left(\frac{\frac{1}{4}(a_1+2a_2+a_3)+b_2T}{\frac{1}{4}(\theta_1+2\theta_2+\theta_3)}-\frac{\frac{1}{4}(b_1+2b_2+b_3)}{\frac{1}{4}(\theta_1^2+2\theta_2^2+\theta_3^2)}\right)$$

$$\left(\frac{e^{-(\frac{1}{4}(\theta_1+2\theta_2+\theta_3)+\frac{1}{4}(k_1+2k_2+k_3))t_1}}{\frac{1}{4}(\theta_1+2\theta_2+\theta_3)+\frac{1}{4}(k_1+2k_2+k_3)}-\frac{e^{-(\frac{1}{4}(\theta_1+2\theta_2+\theta_3)+\frac{1}{4}(k_1+2k_2+k_3))T}}{\frac{1}{4}(\theta_1+2\theta_2+\theta_3)+\frac{1}{4}(k_1+2k_2+k_3)}\right)$$

$$-\left(\frac{\frac{1}{4}(a_1+2a_2+a_3)}{\frac{1}{4}(\theta_1+2\theta_2+\theta_3)}-\frac{\frac{1}{4}(b_1+2b_2+b_3)}{\frac{1}{4}(\theta_1^2+2\theta_2^2+\theta_3^2)}\right)$$

$$\left(\frac{e^{-\frac{1}{4}(k_1+2k_2+k_3)t_1}}{\frac{1}{4}(k_1+2k_2+k_3)}-\frac{e^{-\frac{1}{4}(k_1+2k_2+k_3)T}}{\frac{1}{4}(k_1+2k_2+k_3)}\right)+\frac{\frac{1}{4}(b_1+2b_2+b_3)}{\frac{1}{4}(\theta_1+2\theta_2+\theta_3)}$$

$$\left(\frac{Te^{-\frac{1}{4}(k_1+2k_2+k_3)T}-t_1e^{-\frac{1}{4}(k_1+2k_2+k_3)t_1}}{\frac{1}{4}(k_1+2k_2+k_3)}\right.$$

$$\left.\left.+\frac{e^{-\frac{1}{4}(k_1+2k_2+k_3)T}-e^{-\frac{1}{4}(k_1+2k_2+k_3)t_1}}{\frac{1}{4}(k_1^2+2k_2^2+k_3^2)}\right)\right]+\frac{q^2}{T}r_1m(1-\alpha)+\frac{q}{T}r_2m(1-\alpha)$$

$$I(\tilde{Q})=\frac{1}{4}(Q_1+2Q_2+Q_3)=m(1-\alpha)\frac{D_1+2D_2+D_3}{4}T$$

Now, from Equation (5.4), the membership function of the fuzzy objective function and fuzzy order quantity under the cloudy fuzzy model are given by

$$\gamma_1(Z,T)=\begin{cases}0, & \text{if } Z<Z_{11} \text{ and } Z>Z_{21}\\ \frac{Z-Z_{11}}{Z_{21}-Z_{11}}, & \text{if } Z_{11}\le Z\le Z_{21}\\ \frac{Z_{31}-Z}{Z_{31}-Z_{21}}, & \text{if } Z_{21}\le Z\le Z_{31}\end{cases} \qquad (5.12)$$

where

$$Z_{11} = \frac{qA_0}{T}\left(1 + k(1 - \frac{\rho}{1+T})t\right) + \frac{q}{T}(h_1(1-\alpha) + h_2\alpha)$$

$$\left[\left(\frac{m(1-\alpha) - a(1 - \frac{\rho}{1+T})}{\theta(1 - \frac{\rho}{1+T})} + \frac{b(1 - \frac{\rho}{1+T})}{\theta^2(1 - \frac{\rho}{1+T})}\right)\cdot\right.$$

$$\left(\frac{e^{-(\theta(1-\frac{\rho}{1+T})+k(1-\frac{\rho}{1+T}))t_1} - 1}{\theta(1 - \frac{\rho}{1+T}) + k(1 - \frac{\rho}{1+T})} - \frac{e^{-k(1-\frac{\rho}{1+T})t_1} - 1}{k(1 - \frac{\rho}{1+T})}\right)$$

$$+\frac{b(1 - \frac{\rho}{1+T})}{\theta(1 - \frac{\rho}{1+T})}\left(\frac{t_1 e^{-k(1-\frac{\rho}{1+T})t_1}}{k(1 - \frac{\rho}{1+T})} + \frac{e^{-k(1-\frac{\rho}{1+T})t_1} - 1}{k(1 - \frac{\rho}{1+T})}\right)$$

$$+\left(\frac{a(1 - \frac{\rho}{1+T}) + b(1 - \frac{\rho}{1+T})T}{\theta(1 - \frac{\rho}{1+T})} - \frac{b(1 - \frac{\rho}{1+T})}{\theta^2(1 - \frac{\rho}{1+T})}\right)$$

$$\left(\frac{e^{-(\theta(1-\frac{\rho}{1+T})+k(1-\frac{\rho}{1+T}))t_1}}{\theta(1 - \frac{\rho}{1+T}) + k(1 - \frac{\rho}{1+T})} - \frac{e^{-(\theta(1-\frac{\rho}{1+T})+k(1-\frac{\rho}{1+T}))T}}{\theta(1 - \frac{\rho}{1+T}) + k(1 - \frac{\rho}{1+T})}\right)$$

$$-\left(\frac{a(1 - \frac{\rho}{1+T})}{\theta(1 - \frac{\rho}{1+T})} - \frac{b(1 - \frac{\rho}{1+T})}{\theta^2(1 - \frac{\rho}{1+T})}\right)\left(\frac{e^{-k(1-\frac{\rho}{1+T})t_1}}{k(1 - \frac{\rho}{1+T})} - \frac{e^{-k(1-\frac{\rho}{1+T})T}}{k(1 - \frac{\rho}{1+T})}\right)$$

$$+\frac{b(1 - \frac{\rho}{1+T})}{\theta(1 - \frac{\rho}{1+T})}\left(\frac{Te^{-k(1-\frac{\rho}{1+T})T} - t_1 e^{-k(1-\frac{\rho}{1+T})t_1}}{k(1 - \frac{\rho}{1+T})}\right.$$

$$\left.+\frac{e^{-k(1-\frac{\rho}{1+T})T} - e^{-k(1-\frac{\rho}{1+T})t_1}}{k^2(1 - \frac{\rho}{1+T})}\right) + \frac{q}{T}c\theta\left(1 - \frac{\rho}{1+T}\right)\left[\frac{m(1-\alpha) - a(1 - \frac{\rho}{1+T})}{k(1 - \frac{\rho}{1+T})}\right.$$

$$\left(1 - e^{-k(1-\frac{\rho}{1+T})t_1}\right) + b\left(1 - \frac{\rho}{1+T}\right)\left(\frac{t_1 e^{-k(1-\frac{\rho}{1+T})t_1}}{k(1 - \frac{\rho}{1+T})} + \frac{e^{-k(1-\frac{\rho}{1+T})t_1} - 1}{k^2(1 - \frac{\rho}{1+T})}\right)$$

$$+\left(\frac{a(1 - \frac{\rho}{1+T}) + b(1 - \frac{\rho}{1+T})T}{\theta(1 - \frac{\rho}{1+T})} - \frac{b(1 - \frac{\rho}{1+T})}{\theta^2(1 - \frac{\rho}{1+T})}\right)\left(\frac{e^{-(\theta(1-\frac{\rho}{1+T})+k(1-\frac{\rho}{1+T}))t_1}}{\theta(1 - \frac{\rho}{1+T}) + k(1 - \frac{\rho}{1+T})}\right.$$

$$\left.-\frac{e^{-(\theta(1-\frac{\rho}{1+T})+k(1-\frac{\rho}{1+T}))T}}{\theta(1 - \frac{\rho}{1+T}) + k(1 - \frac{\rho}{1+T})}\right) - \left(\frac{a(1 - \frac{\rho}{1+T})}{\theta(1 - \frac{\rho}{1+T})} - \frac{b(1 - \frac{\rho}{1+T})}{\theta^2(1 - \frac{\rho}{1+T})}\right)$$

$$\left(\frac{e^{-k(1-\frac{\rho}{1+T})t_1}}{k(1 - \frac{\rho}{1+T})} - \frac{e^{-k(1-\frac{\rho}{1+T})T}}{k(1 - \frac{\rho}{1+T})}\right) + \frac{b(1 - \frac{\rho}{1+T})}{\theta(1 - \frac{\rho}{1+T})}$$

$$\left.\left(\frac{Te^{-k(1-\frac{\rho}{1+T})T} - t_1 e^{-k(1-\frac{\rho}{1+T})t_1}}{k(1 - \frac{\rho}{1+T})} + \frac{e^{-k(1-\frac{\rho}{1+T})T} - e^{-k(1-\frac{\rho}{1+T})t_1}}{k^2(1 - \frac{\rho}{1+T})}\right)\right]$$

$$+\frac{q^2}{T}r_1 m(1-\alpha) + \frac{q}{T}r_2 m(1-\alpha),$$

$$Z_{21} = \frac{qA_0}{T}(1+kt) + \frac{q}{T}(h_1(1-\alpha) + h_2\alpha)$$

$$\left[\left(\frac{m(1-\alpha)-a}{\theta} + \frac{b}{\theta^2}\right)\left(\frac{e^{-(\theta+k)t_1}-1}{\theta+k} - \frac{e^{-kt_1}-1}{k}\right)\right.$$

$$+ \frac{b}{\theta}\left(\frac{t_1 e^{-kt_1}}{k} + \frac{e^{-kt_1}-1}{k}\right) + \left(\frac{a+bT}{\theta} - \frac{b}{\theta^2}\right)\left(\frac{e^{-(\theta+k)t_1}}{\theta+k} - \frac{e^{-(\theta+k)T}}{\theta+k}\right)$$

$$- \left(\frac{a}{\theta} - \frac{b}{\theta^2}\right)\left(\frac{e^{-kt_1}}{k} - \frac{e^{-kT}}{k}\right) + \frac{b}{\theta}\left(\frac{Te^{-kT}-t_1 e^{-kt_1}}{k} + \frac{q}{T}c\theta\right)$$

$$\left[\frac{m(1-\alpha)-a}{k}\left(1-e^{-kt_1}\right) + b\left(\frac{t_1 e^{-kt_1}}{k} + \frac{e^{-kt_1}-1}{k^2}\right) + \left(\frac{a+bT}{\theta} - \frac{b}{\theta^2}\right)\right.$$

$$\left(\frac{e^{-(\theta+k)t_1}}{\theta+k} - \frac{e^{-(\theta+k)T}}{\theta+k}\right) - \left(\frac{a}{\theta} - \frac{b}{\theta^2}\right)\left(\frac{e^{-kt_1}}{k} - \frac{e^{-kT}}{k}\right)$$

$$\left. + \frac{b}{\theta}\left(\frac{Te^{-kT}-t_1 e^{-kt_1}}{k} + \frac{e^{-kT}-e^{-kt_1}}{k^2}\right)\right] + \frac{q^2}{T}r_1 m(1-\alpha) + \frac{q}{T}r_2 m(1-\alpha),$$

$$Z_{31} = \frac{qA_0}{T}\left(1+k\left(1+\frac{\sigma}{1+T}\right)t\right) + \frac{q}{T}(h_1(1-\alpha) + h_2\alpha)$$

$$\left[\left(\frac{m(1-\alpha)-a(1+\frac{\sigma}{1+T})}{\theta(1+\frac{\sigma}{1+T})} + \frac{b(1+\frac{\sigma}{1+T})}{\theta^2(1+\frac{\sigma}{1+T})}\right)\right]$$

$$\left(\frac{e^{-(\theta(1+\frac{\sigma}{1+T})+k(1+\frac{\sigma}{1+T}))t_1}-1}{\theta(1+\frac{\sigma}{1+T})+k(1+\frac{\sigma}{1+T})} - \frac{e^{-k(1+\frac{\sigma}{1+T})t_1}-1}{k(1+\frac{\sigma}{1+T})}\right) + \frac{b(1+\frac{\sigma}{1+T})}{\theta(1+\frac{\sigma}{1+T})}$$

$$\left(\frac{t_1 e^{-k(1+\frac{\sigma}{1+T})t_1}}{k(1+\frac{\sigma}{1+T})} + \frac{e^{-k(1+\frac{\sigma}{1+T})t_1}-1}{k(1+\frac{\sigma}{1+T})}\right)$$

$$+ \left(\frac{a(1+\frac{\sigma}{1+T})+b(1+\frac{\sigma}{1+T})T}{\theta(1+\frac{\sigma}{1+T})} - \frac{b(1+\frac{\sigma}{1+T})}{\theta^2(1+\frac{\sigma}{1+T})}\right)$$

$$\left(\frac{e^{-(\theta(1+\frac{\sigma}{1+T})+k(1+\frac{\sigma}{1+T}))t_1}}{\theta(1+\frac{\sigma}{1+T})+k(1+\frac{\sigma}{1+T})} - \frac{e^{-(\theta(1+\frac{\sigma}{1+T})+k(1+\frac{\sigma}{1+T}))T}}{\theta(1+\frac{\sigma}{1+T})+k(1+\frac{\sigma}{1+T})}\right)$$

$$- \left(\frac{a(1+\frac{\sigma}{1+T})}{\theta(1+\frac{\sigma}{1+T})} - \frac{b(1+\frac{\sigma}{1+T})}{\theta^2(1+\frac{\sigma}{1+T})}\right)\left(\frac{e^{-k(1+\frac{\sigma}{1+T})t_1}}{k(1+\frac{\sigma}{1+T})} - \frac{e^{-k(1+\frac{\sigma}{1+T})T}}{k(1+\frac{\sigma}{1+T})}\right)$$

$$+ \frac{b(1+\frac{\sigma}{1+T})}{\theta(1+\frac{\sigma}{1+T})}\left(\frac{Te^{-k(1+\frac{\sigma}{1+T})T}-t_1 e^{-k(1+\frac{\sigma}{1+T})t_1}}{k(1+\frac{\sigma}{1+T})}\right.$$

$$+ \frac{e^{-k(1+\frac{\sigma}{1+T})T}-e^{-k(1+\frac{\sigma}{1+T})t_1}}{k^2 k(1+\frac{\sigma}{1+T})}\right) + \frac{q}{T}c\theta\left(1+\frac{\sigma}{1+T}\right)\left[\frac{m(1-\alpha)-a(1+\frac{\sigma}{1+T})}{k(1+\frac{\sigma}{1+T})}\right.$$

$$\left. \left(1-e^{-k(1+\frac{\sigma}{1+T})t_1}\right) + b\left(1+\frac{\sigma}{1+T}\right)\left(\frac{t_1 e^{-k(1+\frac{\sigma}{1+T})t_1}}{k(1+\frac{\sigma}{1+T})} + \frac{e^{-k(1+\frac{\sigma}{1+T})t_1}-1}{k^2(1+\frac{\sigma}{1+T})}\right)\right.$$

$$+\left(\frac{a(1+\frac{\sigma}{1+T})+b(1+\frac{\sigma}{1+T})T}{\theta(1+\frac{\sigma}{1+T})}-\frac{b(1+\frac{\sigma}{1+T})}{\theta^2(1+\frac{\sigma}{1+T})}\right)\left(\frac{e^{-(\theta(1+\frac{\sigma}{1+T})+k(1+\frac{\sigma}{1+T}))t_1}}{\theta(1+\frac{\sigma}{1+T})+k(1+\frac{\sigma}{1+T})}\right.$$

$$-\frac{e^{-(\theta(1+\frac{\sigma}{1+T})+k(1+\frac{\sigma}{1+T}))T}}{\theta(1+\frac{\sigma}{1+T})+k(1+\frac{\sigma}{1+T})}\right)-\left(\frac{a(1+\frac{\sigma}{1+T})}{\theta(1+\frac{\sigma}{1+T})}-\frac{b(1+\frac{\sigma}{1+T})}{\theta^2(1+\frac{\sigma}{1+T})}\right)$$

$$\left(\frac{e^{-k(1+\frac{\sigma}{1+T})t_1}}{k(1+\frac{\sigma}{1+T})}-\frac{e^{-k(1+\frac{\sigma}{1+T})T}}{k(1+\frac{\sigma}{1+T})}\right)+\frac{b(1+\frac{\sigma}{1+T})}{\theta(1+\frac{\sigma}{1+T})}$$

$$\left(\frac{Te^{-k(1+\frac{\sigma}{1+T})T}-t_1e^{-k(1+\frac{\sigma}{1+T})t_1}}{k(1+\frac{\sigma}{1+T})}+\frac{e^{-k(1+\frac{\sigma}{1+T})T}-e^{-k(1+\frac{\sigma}{1+T})t_1}}{k^2(1+\frac{\sigma}{1+T})}\right)\Bigg]$$

$$+\frac{q^2}{T}r_1m(1-\alpha)+\frac{q}{T}r_2m(1-\alpha),$$

and

$$\gamma_2(Q,\,T)=\begin{cases}0,\;\;\text{if } Q < Q_{11} \text{ and } Q > Q_{21}\\\frac{Q-Q_{11}}{Q_{21}-Q_{11}},\;\;\text{if } Q_{11} \le Q \le Q_{21}\\\frac{Q_{31}-Q}{Q_{31}-Q_{21}},\;\;\text{if } Q_{21} \le Q \le Q_{31}\end{cases}\qquad(5.13)$$

where

$$\begin{cases}Q_{11}=m\,(1-\alpha)\,D\left(1-\frac{\rho}{1+T}\right)T\\Q_{21}=m\,(1-\alpha)\,DT\\Q_{11}=m\,(1-\alpha)\,D\left(1+\frac{\sigma}{1+T}\right)T\end{cases}$$

Now, from Equation (5.5), the index values of cloudy fuzzy objective and cloudy fuzzy order quantity are respectively given by

$$I(\tilde{Z})=\frac{1}{\beta}\int_{T=0}^{\beta}\frac{1}{4}(Z_{11}+2Z_{21}+Z_{31})dT$$

$$=\frac{1}{4\beta}\int_{T=0}^{\beta}(Z_{11}+2Z_{21}+Z_{31})dT$$

[**Remark:** The above expression leads to a very large mathematical representation and tedious work too. For that reason, only the value is calculated with the help of a numerical example. To avoid the unethical value, a small number of ε is considered where the value of ε is $\to 0$.]

$$I(\tilde{Q})=\frac{1}{\beta}\int_{T=0}^{\beta}\frac{1}{4}(Q_{11}+2Q_{21}+Q_{31})dT$$

$$= \frac{1}{4\beta} \int_{T=0}^{\beta} (Q_{11} + 2Q_{21} + Q_{31}) dT$$

$$= \frac{m(1-\alpha)}{4\beta} [2\beta^2 + (\sigma - \rho)(\beta - \log(1+\beta))]$$

5.6 Numerical Example

Let us assume that A_0 = \$500/order, h_1 = \$300/unit, α = 0.9, h_2 = \$400/unit, m = 10, c = \$1000/unit, r_1 = \$50/unit, r_2 = \$200/unit, a = 30 units and b = 40 units for crisp model and $<a_1, a_2, a_3>$ = $<25, 30, 40>$, $<b_1, b_2, b_3>$ = $<35, 40, 50>$ for fuzzy model, θ = 0.05 for crisp model and $<\theta_1, \theta_2, \theta_3>$ = $<0.03, 0.05, 0.06>$ for fuzzy model, k = 0.6 for crisp model and $<k_1, k_2, k_3>$ = $<0.4, 0.6, 0.7>$ for fuzzy model. The values of σ, ρ and ε for cloudy fuzzy model are considered as σ = 0.16, ρ = 0.19 and ε = 0.0001 and the obtained results from Equations (5.9–5.13) are shown in Table 5.1. The degree of fuzziness is calculated from the formula $d_f = \frac{U_b - L_b}{2m_1}$ where U_b and L_b are the upper and lower bounds of fuzzy demand and fuzzy deterioration respectively, and m_1 is their corresponding mode. Assuming t = 10, the fuzzy demand and deterioration becomes $<375, 430, 540>$ and $<0.03, 0.05, 0.06>$, respectively. Here, the corresponding mean, median and mode for fuzzy demand is calculated as 448.33, 430 and 393.34, respectively and similarly, mean, median and mode for fuzzy deterioration is calculated as 0.04667, 0.05 and 0.05666, respectively, where the mode (m_1) = 3× median − 2 ×mean.

Table 5.1 Computational results

Model	Cycle Time T	Inventory Period t_1	Order Quantity Q	Total Cost Z	Cloud Index CI	Degree of Fuzziness d_f
Crisp Model	5.47	4.59	438.07	1132.21	...	...
Fuzzy Model	5.24	4.17	451.36	1076.37	...	0.419 (fuzzy demand), 0.529 (fuzzy deterioration)
Cloudy Fuzzy Model	10.20	8.11	525.72	875.14	0.237	...

5.7 Sensitivity Analysis

We now study the effects of changes in the system parameters a, b, k, θ, α, r_1, h_1, and c on the optimal values of T, t_1, Q and total cost Z for both the crisp, fuzzy and cloudy fuzzy model. This sensitivity analysis is performed by changing each of the parameters by $+30\%$, $+10\%$, -10% and -30%, taking one parameter at a time and keep the remaining parameters unchanged. The results based on the above example are shown in Tables 5.2, 5.3, and 5.4, and on the basis of these results, Figure 5.2 is drawn.

The following observations are made from Tables 5.2, 5.3, and 5.4, which will clarify the importance of our model.

a) When the parameters of the demand constraint a and b increase, then the total cost of the system decreases for the crisp, fuzzy and cloudy fuzzy model but from the tables, one can observe that the total cost is minimum for the cloudy fuzzy model with respect to the crisp and fuzzy model.

b) When the rate of inflation decreases, the total cost of the system decreases for the crisp, fuzzy and cloudy fuzzy model but the total cost is minimum for the cloudy fuzzy model rather than a crisp and fuzzy model. From the analysis, it is shown that uncertainties in inflation rate are more practical than the constant inflation rate.

c) When the rate of deterioration increases, the holding cost for both defective items h_1 and non-defective items h_2 increase, which will expand the total cost of the system inevitably.

d) A higher percentage of defective items α in the system will raise the inspection cost r_1 of the system during each production cycle. At the same time, the number of shipments for repairing of damageable items from the retailer to the supplier per lot becomes more, and also shipment size increases, which will cause damage in the amount for the retailer.

e) When the unit purchasing cost c increases, the total cost of the system increases and, consequently, the total profit for the retailer decreases. So, the manager should order a greater quantity at a time to avoid a higher charge against purchasing cost, if the deterioration cost will be less than the purchasing cost.

5.8 Concluding Remarks and Future Studies

In this chapter, we have formulated an inventory model with fuzzy demand, fuzzy deterioration rate, and fuzzy inflation rate and we have compared the result through NGTFN and CNTFN. Since the demand of many items,

Table 5.2 Sensitivity analysis for various parameters for crisp model

Parameters	% Change	Values	Cycle Time T	Inventory Period t_1	Order Quantity Q	Total Cost Z
a	+30	39	5.46	4.68	435.11	1021.57
	+10	33	5.46	4.68	433.10	1045.21
	−10	27	5.44	4.68	432.27	1068.09
	−30	21	5.44	4.68	430.00	1094.42
b	+30	52	5.41	4.65	440.51	1047.96
	+10	44	5.40	4.64	449.36	1085.68
	−10	36	5.39	4.63	448.08	1132.74
	−30	28	5.39	4.60	446.47	1176.08
k	+30	0.78	5.31	4.55	452.35	1109.51
	+10	0.66	5.28	4.54	450.22	1127.23
	−10	0.54	5.22	4.53	448.09	1164.02
	−30	0.42	5.17	4.53	448.00	1183.37
θ	+30	0.065	5.44	4.63	444.39	1399.90
	+10	0.055	5.40	4.61	443.94	1371.28
	−10	0.045	5.37	4.59	442.37	1293.77
	−30	0.035	5.35	4.56	441.74	1218.00
α	+30	1.17	5.56	4.64	443.28	1385.23
	+10	0.99	5.54	4.63	442.15	1350.68
	−10	0.81	5.53	4.61	441.07	1287.49
	−30	0.63	5.51	4.58	440.64	1242.16
r_1	+30	65	5.52	4.65	435.72	1267.56
	+10	55	5.51	4.63	433.53	1232.00
	−10	45	5.48	4.61	431.60	1188.07
	−30	35	5.45	4.59	430.25	1161.10
h_2	+30	390	5.57	4.44	455.21	1568.87
	+10	330	5.56	4.42	453.37	1539.26
	−10	270	5.54	4.40	451.64	1477.80
	−30	210	5.51	4.37	450.71	1456.19
c	+30	1300	5.48	4.53	444.28	1483.59
	+10	1100	5.44	4.52	442.17	1452.07
	−10	900	5.41	4.50	441.06	1407.34
	−30	700	5.37	4.50	440.59	1394.55

like seasonal foods, clothes, are strongly dependent on time, therefore, the demand is regarded as time-dependent instead of constant demand, and then fuzzy demand. For delivering a highly demandable product, the company has to produce a large amount of items within a very short period of time. So, it is very obvious that the production process will be imperfect and the company

Table 5.3   Sensitivity analysis for various parameters for fuzzy model

Parameters	% Change	Values	Fuzzy Model Cycle Time T	Inventory Period t_1	Order Quantity Q	Total Cost Z
a	+30	39	5.33	4.28	447.23	1142.00
	+10	33	5.32	4.27	446.79	1276.53
	−10	27	5.31	4.27	446.32	1297.31
	−30	21	5.30	4.26	446.10	1305.24
b	+30	52	5.32	4.27	455.21	1198.47
	+10	44	5.32	4.26	455.19	1265.08
	−10	36	5.30	4.25	455.15	1294.12
	−30	28	5.30	4.25	455.11	1348.30
k	+30	0.78	5.26	4.19	452.73	1448.19
	+10	0.66	5.26	4.18	451.69	1407.43
	−10	0.54	5.26	4.18	450.85	1342.36
	−30	0.42	5.25	4.18	449.30	1215.04
θ	+30	0.065	5.10	4.20	443.25	1372.15
	+10	0.055	5.07	4.20	443.20	1254.83
	−10	0.045	5.07	4.20	443.16	1139.74
	−30	0.035	5.07	4.20	443.10	1068.55
α	+30	1.17	5.15	4.22	448.56	1167.38
	+10	0.99	5.14	4.21	447.43	1059.50
	−10	0.81	5.13	4.20	446.39	985.29
	−30	0.63	5.11	4.19	445.35	924.36
r_1	+30	65	5.20	4.25	450.14	1269.28
	+10	55	5.20	4.23	450.27	1147.54
	−10	45	5.19	4.21	450.38	1068.39
	−30	35	5.18	4.20	450.77	1013.70
h_2	+30	390	5.22	4.33	447.34	1576.36
	+10	330	5.21	4.32	445.90	1457.28
	−10	270	5.20	4.31	443.75	1381.17
	−30	210	5.19	4.30	441.56	1245.36
c	+30	1300	5.27	4.28	460.04	1366.28
	+10	1100	5.26	4.27	461.81	1314.75
	−10	900	5.24	4.26	462.35	1238.00
	−30	700	5.21	4.26	463.57	1147.28

will deliver a large amount of imperfect items. Among these imperfect items, some products are less defective and some are repairable at a lower cost; those products will be delivered at a lower price. Some products with defective items are highly damageable; they will be rejected immediately from the market and for that process, an inspection policy has been required and which has included in our model.

Table 5.4 Sensitivity analysis for various parameters for cloudy fuzzy model

Parameters	% Change	Values	Cycle Time T	Inventory Period t_1	Order Quantity Q	Total Cost Z
a	+30	39	11.76	9.70	559.24	870.11
	+10	33	11.38	9.59	550.00	872.39
	−10	27	10.82	9.47	548.59	874.50
	−30	21	10.45	9.33	536.05	876.71
b	+30	52	10.67	9.97	568.31	837.65
	+10	44	10.43	9.73	565.30	840.09
	−10	36	10.28	9.56	563.27	844.51
	−30	28	10.10	9.38	561.69	849.66
k	+30	0.78	9.78	8.79	530.65	921.13
	+10	0.66	9.72	8.52	528.28	919.79
	−10	0.54	9.70	8.31	526.74	917.65
	−30	0.42	9.77	8.16	524.55	914.56
θ	+30	0.065	10.87	9.25	535.04	911.48
	+10	0.055	10.64	9.38	534.36	909.23
	−10	0.045	10.53	9.47	533.51	907.40
	−30	0.035	10.41	9.62	532.64	904.78
α	+30	1.17	11.62	10.00	540.73	891.27
	+10	0.99	11.73	10.16	541.50	878.56
	−10	0.81	11.81	10.24	542.88	873.74
	−30	0.63	11.89	10.38	543.67	867.08
r_1	+30	65	11.50	7.56	548.09	811.35
	+10	55	11.42	7.69	546.13	809.49
	−10	45	11.39	7.73	544.18	807.38
	−30	35	11.31	7.84	542.00	804.74
h_2	+30	390	10.64	8.97	525.89	850.64
	+10	330	10.67	8.83	524.74	847.48
	−10	270	10.75	8.75	523.55	843.77
	−30	210	10.83	8.64	521.37	841.56
c	+30	1300	11.26	9.20	531.26	863.12
	+10	1100	11.49	9.29	530.05	860.45
	−10	900	11.57	9.37	529.33	856.87
	−30	700	11.78	9.41	527.49	852.38

Here, all the inventory models have studied through crisp, NGTFN and CNTFN while the concept of cloudy fuzzy number is quite new in literature. From numerical example and sensitivity analysis, it is seen that values of inventory cycle time are larger but the values of the degree of fuzziness and cloud index are smaller. It is really attracting that the value of total cost

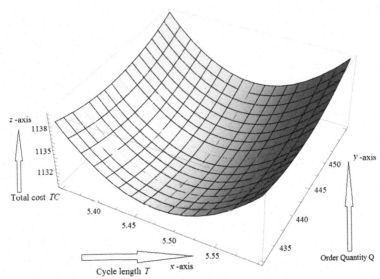

Figure 5.2 Graphical presentation to show the convexity of the total cost for crisp model.

is minimum for a cloudy fuzzy model than the crisp or normal fuzzy model. The global optimality of total cost is validated through a figure. The model will be useful for the decision maker in planning and controlling the industry. The model adds managerial insight which helps industry to obtain a financial surplus at an optimal level.

A possible future study for the proposed model is to consider a multi-item inventory model under the trade-credit policy. One can also add a reliability function (constant or variable), probabilistic demand, variable lead time, etc. The model can also be extended to consider set-up cost reduction in production time, and variable shipment size, too. Further, we can extend the work by collecting some real data from industry. Then we can also allow some more managerial insights based on that collecting real data.

References

[1] Allahviranloo, T. and Saneifard, R. (2012) Defuzzification method for ranking fuzzy numbers based on center of gravity. *Iranian Journal of Fuzzy Systems*, 9(6), 57–67.

[2] Ban, A.I. and Coroianu, L. (2014) Existence, uniqueness and continuity of trapezoidal approximations of fuzzy numbers under a general condition. *Fuzzy Sets and Systems*, 257, 3–22.

[3] Bellman, R.E. and Zadeh, L.A. (1970) Decision-making in a fuzzy environment, *Management Sciences*, 17, B141–B164.

[4] De, S.K. and Goswami, A. (2001) A replenishment policy for items with finite production rate and fuzzy deterioration rate, *Opsearch*, 38(4), 419–430.

[5] De, S.K. and Goswami, A. (2006) An EOQ model with fuzzy inflation rate and fuzzy deterioration rate when a delay in payment is permissible. International Journal of Systems Science, 37(5), 323–335.

[6] De, S.K. and Beg, I. (2016) Triangular dense fuzzy Neutrosophic sets. *Neutrosophic Sets and Systems*, 13, 1–12.

[7] De, S.K. and Mahata, G.C. (2016) Decision of a fuzzy inventory with fuzzy backorder model under cloudy fuzzy demand rate. *International Journal of Applied and Computational Mathematics*, DOI:10.1007/s40819-016-0258-4.

[8] Deng, H. (2014) Comparing and ranking fuzzy numbers using ideal solutions. *Applied Mathematical Modelling*, 38, 1638–1646.

[9] Ezzati, R., Allahviranloo, T., Khezerloo, S. and Khezerloo, M. (2012) An approach for ranking of fuzzy numbers. *Expert Systems with Applications*, 39, 690–695.

[10] Hajjari, T. and Abbasbandy, S. (2011) A note on "The revised method of ranking LR fuzzy number based on deviation degree". *Expert Systems with Applications*, 39, 13491–13492.

[11] Harris, F. (1915) Operations and Cost. Factory Management Service. A.W. Shaw Co, Chicago.

[12] Kao, C. and Hsu, W.K. (2002) Lot size reorder point inventory model with fuzzy demands. *Computers & Mathematics with Applications*, 43, 1291–1302.

[13] Kauffman, A. and Gupta, M.M. (1992) Introduction to Fuzzy Arithmetic Theory and Applications. Van Nostrand Reinhold, New York.

[14] Kazemi, N., Olugu, E.U., Salwa Halim, A.R. and Ghazilla, R.A.B.R. (2015) Development of a fuzzy economic order quantity model for imperfect quality items using the learning effect on fuzzy parameters. *Journal of Intelligent & Fuzzy Systems*, 28(5), 2377–2389.

[15] Mahata, G.C. and Goswami, A. (2007) An EOQ model for deteriorating items under trade credit financing in the fuzzy sense. *Production Planning and Control*, 18, 681–692.

[16] Mahata, G.C. (2015) A production inventory model with imperfect production process and partial backlogging under learning considerations in fuzzy random environments. *Journal of Intelligent Manufacturing*, 28(4), 883–897.

[17] Moon, I. and Lee, S. (2000) The effect of inflation and time value of money on an economic order quantity model with a random product life cycle. *European Journal of Operational Research*, 125, 588–601.

[18] Pervin, M., Mahata, G.C. and Roy, S.K. (2016) An inventory model with demand declining market for deteriorating items under trade credit policy. *International Journal of Management Science and Engineering Management*, 11, 243–251.

[19] Pervin, M., Roy, S.K. and Weber, G.W. (2018) Analysis of inventory control model with shortage under time-dependent demand and time-varying holding cost including stochastic deterioration. *Annals of Operations Research*, 260, 437–460.

[20] Pervin, M., Roy, S.K. and Weber, G.W. (2018) A Two-echelon inventory model with stock-dependent demand and variable holding cost for deteriorating items. *Numerical Algebra, Control and Optimization*, 7(1), 21–50.

[21] Pervin, M., Roy, S.K. and Weber, G.W. (2018) An integrated inventory model with variable holding cost under two levels of trade-credit policy. *Numerical Algebra, Control and Optimization*, 8(2), 169–191.

[22] Pervin, M., Roy, S.K. and Weber, G.W. (2018) Multi-item deteriorating two-echelon inventory model with price- and stock-dependent demand: A trade-credit policy. *Journal of Industrial and Management Optimization*, DOI:10.3934/jimo.2018098.

[23] Pervin, M., Roy, S.K. and Weber, G.W. (2019) Deteriorating inventory with preservation technology under price- and stock-sensitive demand. *Journal of Industrial and Management Optimization*, DOI: 10.3934/jimo.2019019.

[24] Roy, S.K., Pervin, M. and Weber, G.W. (2018) A two-warehouse probabilistic model with price discount on backorders under two levels of trade-credit policy. *Journal of Industrial and Management Optimization*, DOI:10.3934/jimo.2018167.

[25] Skouri, K. and Papachristos, S. (2002) Note on deterministic inventory lot-size models under inflation with shortages and deterioration for fluctuating demand by Yang et al. *Naval Research Logistics*, 49, 527–529.

[26] Vahidian, A. and Tareghian, H.R. (1999) Production planning in fuzzy environment. *The Korean Journal of Computational & Applied Mathematics*, 6(2), 315–330.

[27] Wang, Z.X., Liu, Y.J., Fan, Z.P. and Feng, B. (2009) Ranking L-R fuzzy number based on deviation degree. *Information Sciences*, 179, 2070–2077.

[28] Wu, D.D., Zhang, Y., Wu, D. and Olson, D.L. (2010) Fuzzy multi-objective programming for supplier selection and risk modeling: A possibility approach. *European Journal of Operational Research*, 200, 774–787.

[29] Xu, P., Su, X., Wu, J., Sun, X., Zhang, Y. and Deng, Y. (2012) A note on ranking generalized fuzzy numbers. *Expert Systems with Applications*, 39, 6454–6457.

[30] Yager, R.R. (1981) A procedure for ordering fuzzy subsets of the unit interval, *Information Sciences*, 24, 143–161.

[31] Yang, H.L., Teng, J.T. and Chern, M.S. (2010) An inventory model under inflation for deteriorating items with stock-dependent consumption rate and partial backlogging shortages. *International Journal of Production Economics*, 123(1), 8–19.

[32] Zadeh, L.A. (1965), Fuzzy sets, *Information and Control*, 8, 338–353.

6

Summability and Its Application for the Stability of the System

Smita Sonker and Alka Munjal[*]

National Institute of Technology Kurukshetra, Haryana, India
E-mail: smita.sonker@gmail.com; alkamunjal8@gmail.com
*Corresponding Author

Summability methods are a useful tool in dealing with the non-convergent series/integrals and assigns a value (number) to it. Signals can be in the form of various types of series (Infinite Series, Fourier series etc.) and hence, summability theory is applicable in finding the error of approximation. In this chapter, the authors gave an introductory discussion on summability process, regularity of summability process and it's Silverman-Toeplitz theorem. Further, they explained about the stability of the frequency response of the system.

6.1 Introduction

In mathematics, a series is the sum of the terms of an infinite sequence of numbers. An infinite sequence $\{a_1, a_2, a_3, \ldots\}$, the nth partial sum s_n, is the sum of the first n terms of the sequence, that is,

$$s_n = \sum_{k=1}^{n} a_k$$

A series is **convergent** if the sequence of its partial sums $\{s_1, s_2, s_3, \ldots\}$, tends to a limit; that means that the partial sums become closer and closer to a given number when the number of their terms increases. More precisely, a series converges, if there exists a number l such that for any arbitrarily small positive number ε, there is a (sufficiently large) integer N such that for all $n \geq N$,

$$|s_n - l| \le \varepsilon$$

If the series is convergent, the number l (necessarily unique) is called the **sum of the series**.

Any series that is not convergent is said to be divergent.

In mathematics, a **divergent series** is an infinite series that is not convergent, meaning that the infinite sequence of the partial sums of the series does not have a finite limit.

If a series converges, the individual terms of the series must approach zero. Thus, any series in which the individual terms do not approach zero diverges. However, convergence is a stronger condition: not all series whose terms approach zero converge.

In specialized mathematical contexts, values can be objectively assigned to certain series whose sequences of partial sums diverge, in order to make meaning of the divergence of the series. A **summability method** or **summation method** is a partial function from the set of series to values.

For example, Cesàro summation assigns Grandi's divergent series

$$1 - 1 + 1 - 1 + 1 - 1 \dots$$

the value 1/2. Cesàro summation is an **averaging** method, in that it relies on the arithmetic mean of the sequence of partial sums. Other methods involve analytic continuations of related series. In physics, there are a wide variety of summability methods; these are discussed in greater detail in the article on regularization.

The theory of convergence and summation of series is not always a classical section of the analysis. For a long time, the idea that such infinite summation could produce a finite result was considered paradoxical by many mathematicians and philosophers. This paradox was resolved using the concept of a limit during the 19th century. Greek mathematician Archimedes (c.287BC-c.212BC) produced the first known summation of an infinite series with a method that is still used in the area of calculus. He used the method of exhaustion to calculate the area under the arc of a parabola with the summation of an infinite series.

In modern terminology, a "sequence" (called a "progression" in British English) is an ordered list of numbers; the numbers in this ordered list are called the "elements" or the "terms" of the sequence. A "series" is adding up of all the terms of a sequence; the addition, and the resulting value, are called the "sum" or the "summation". A series which is in the form of $a_1 + a_2 + a_3 + \dots = \sum_{n=1}^{\infty} a_n$ is called an infinite series, where $\{a_n\}$ is an infinite sequence.

When the limit of $\sum a_n$ exists, then the series is convergent or summable. If an infinite series $\sum a_n$ is not summable in some sense and the series $\sum a_n \lambda_n$ is summable within the same sense, then λ_n is said to be a summability factor of the infinite series $\sum a_n$. If the summability is absolute, the factors are called absolute summability factors.

6.2 Summability Process

The theory of convergence of series/integrals came into existence on the solid foundations of two splendid researches viz. Cauchy's Analyze Algébrique in 1821 [1] and Abel's researches on Binomial series in 1826 [2]. But it was still not possible to deal satisfactorily with series having oscillating sequences of partial sums. However, these non-convergent series were quite useful and various operations performed on them often led to important results which could verified independently. Endeavors of some eminent mathematicians made it possible to develop satisfactory methods to associate processes closely connected with Cauchy's concept of convergence. Moreover, various extensions (summability methods) of the classical concept of convergence (contemporary theory of summability) for divergent series/improper integrals came into light towards the end of the 19th century.

Summability is a field in which we study the non-convergent series/integrals and assigns a value (number) to it. In mathematical analysis, summability method is an alternative formulation of convergence of a series which is divergent in the conventional sense. Some values can be objectively assigned to certain series whose sequence of partial sums diverges. In 1890 [3], Cesàro was the first to deal with the sum of divergent series and defined **Cesàro summation**.

For instance, followings are some interesting cases, for which the assignment of limits can be done,

- Real or complex sequences for the limit process '$n \rightarrow \infty$',
- Series (convergence of series),
- Sequences and series of functions like power series, Fourier series, etc.,
- Limit of a function at a point (continuity, continuous extension),
- Differentiation of functions,
- Integration of functions.

For example, Cesàro summation assigns Grandi's series $1-1+1-1+1- \cdots$ the value $1/2$.

Throughout the 19th century, many mathematicians studied various divergent series and defined numerous Summability methods namely Abel summability, Cesàro summability, Euler summability, Hausdorff summability, Nörlund summability, Riesz summability etc. Toeplitz (1881–1940) presented the summability methods in a most appropriate way. The summability theory has been also developed by Bosanquet [4, 5], Hyslop [6], Chow [7], Wang [8], Bor [9, 10], Özarslan [11], Özarslan and Ari [12], Sulaiman [13, 14] and other mathematician, particularly with regard to its use in the study of series.

6.3 Types of Summability

There exist three types of summability,

> Ordinary Summability,
> Strong Summability,
> Absolute Summability.

6.3.1 Ordinary Summability

Let $\sum a_n$ be an infinite series of real numbers with a sequence of partial sums $\{s_n\}$ and $T = (a_{n,k})$ be an infinite matrix with real or complex constants.

$$
T = \begin{bmatrix}
a_{11} & a_{12} & \cdots & a_{1n} & \cdots \\
a_{21} & a_{22} & \cdots & a_{2n} & \cdots \\
\cdots & \cdots & \cdots & \cdots & \cdots \\
a_{m1} & a_{m2} & \cdots & a_{mn} & \cdots \\
\cdots & \cdots & \cdots & \cdots & \cdots
\end{bmatrix}
$$

Then, the sequence-to-sequence transformation $t_n = \sum_{k=1}^{n} a_{n,k} s_k$ defines the matrix transform of the sequence $\{s_n\}_{n=1}^{\infty}$. The sequence $\{s_n\}$ or the series $\sum a_n$ is said to be matrix summable to s, if $\lim_{n \to \infty} t_n = s$.

6.3.2 Strong Summability

The infinite series $\sum_{n=0}^{\infty} a_n$ with the sequence of the partial sum $\{s_n\}$ is strong summable with index k by the method A (A-summable) to the limit s, if it is A-summable to s,

and
$$
\sum_{i=1}^{n} i^k |t_i - t_{i-1}|^k = O(n), \quad \text{as } n \to \infty, \tag{6.1}
$$

It is denoted by $[A]_k$ or $[A, \; k]$.

6.3.3 Absolute Summability

The infinite series $\sum_{n=0}^{\infty} a_n$ with the sequence of the partial sum $\{s_n\}$ is absolute summable by the method A (A-summable) to the limit s, if it is A-summable to s and the sequence $\{t_n\}$ is of bounded variation, i.e. mathematically

$$\lim_{n \to \infty} t_n = s \qquad (6.2)$$

and

$$\sum_{n=1}^{\infty} |t_n - t_{n-1}| < \infty. \qquad (6.3)$$

The infinite series $\sum_{n=0}^{\infty} a_n$ with the sequence of the partial sum $\{s_n\}$ is absolute summable with index k by the method A (A-summable) to the limit s, if it is A-summable to s and the sequence $\{t_n\}$ is of k indexed bounded variation, i.e. mathematically

$$\lim_{n \to \infty} t_n = s$$

and

$$\sum_{n=1}^{\infty} n^{k-1} |t_n - t_{n-1}|^k < \infty. \qquad (6.4)$$

It is denoted by $|A|_k$ or $|A, k|$.

6.4 Regularity of a Summability Process

The summability matrix T or the sequence-to-sequence transformation t_k is said to be regular, if

$$\lim_{n \to \infty} s_n = s \quad \Leftrightarrow \quad \lim_{n \to \infty} t_n = s.$$

Silverman and Toeplitz gave a theorem defining the necessary and sufficient conditions for the summability method to be regular which further known as **Silverman-Toeplitz Theorem**.

6.5 Silverman-Toeplitz Theorem

In order that T should be regular, it is necessary and sufficient that

$\sum_k |a_{n,k}| \leq M$, where M is a finite constant independent of n,

$$\lim_{n \to \infty} a_{n,k} = 0, \quad \text{for each } k,$$

and $\quad \lim_{n \to \infty} \sum_{k=0}^{n} a_{n,k} = 1.$

The regular matrices are called Toeplitz matrices.

6.6 Application of Absolute Summable Factor for Stability

The purpose of applying absolute summability is to ensure control over the output so that the system remains stable for the assumed conditions. This process will make the system stable if the input data satisfy the conditions which are required for the absolute summability. Consider, the input (load) data is in the form of a periodic signal that can be represented as a harmonic summation of sinusoids. The input data is unbounded and hence the resulted variation in the output data would not be represented as a bounded variation. Summability methods are suitable for converting the unbounded data into a bounded data or filtering the data for stability.

A necessary and sufficient condition for a system to be BIBO (Bounded Input Bounded Output) stable is that the impulse response be absolutely summable, i.e.,

$$\text{BIBO stable} \iff \sum_{n=-\infty}^{\infty} |h(n)| < \infty.$$

Summability techniques are trained to minimize the error. With the use of summability technique, the output of the signals can be made stable, bounded and used to predict the behavior of the input data, the initial situation and the changes in the complete process.

The absolute summable factor has been applied to an infinite series, in order to obtain the frequency response, frequency response function and behaviour of magnitude of frequency response of the system. The details of the investigation are as follows.

6.6.1 Stability of the Frequency Response of the Moving Average System

Let there exist a moving average system [15] and its impulse response is

$$I_n = \begin{cases} \dfrac{1}{X_1 + X_2 + 1}, & -X_1 \leq n \leq X_2, \\ 0, & \text{otherwise.} \end{cases}$$

The representation of frequency response (FR) of system is

$$H\left(e^{jw}\right) = \sum_{n=-\infty}^{\infty} I_n e^{-jwn}.$$

The frequency response function (FRF) exists if the condition of stability is satisfied by the system because the condition of stability is sufficient for FRF to exist.

$$\left|H\left(e^{jw}\right)\right| = \left|\sum_{n=-\infty}^{\infty} I_n e^{-jwn}\right|$$
$$\leq \sum_{n=-\infty}^{\infty} \left|I_n e^{-jwn}\right|$$
$$\leq \sum_{n=-\infty}^{\infty} |I_n|.$$

That is, the transient response is bounded by the sum of the absolute values of all the impulse response samples. If the right-hand side is bounded, i.e., if

$$\sum_{n=-\infty}^{\infty} |I_n| < \infty,$$

then the system is stable. It follows that, for stable systems, the transient response must become increasingly smaller as $n \to \infty$. Thus, a sufficient condition for the transient response to decay asymptotically (die out) is that the system be stable.

Thus, if I_n satisfy the condition of absolutely summable, then $H(e^{jw})$ exists. In this condition, the infinite series will uniformly converge to a function of w, which is continuous. A stable sequence is absolutely summable and all stable sequences have Fourier transforms. Hence, in this way stable system, having an absolutely summable impulse response, can made finite and contains continuous frequency response.

$$H\left(e^{jw}\right) = \frac{1}{X_1 + X_2 + 1} \frac{\sin[w(X_1 + X_2 + 1)/2]}{\sin(w/2)} e^{-jw(X_2 - X_1)/2}.$$

Figure 6.1 represent the plot of magnitude of $H(e^{jw})$ for $X_1 = 0$ and $X_2 = 4$ and the typical values of $H(e^{jw})$ up to first ten terms have been reported in Table 6.1. This can be expressed as

$$H\left(e^{jw}\right) = \frac{1}{M_1 + M_2 + 1} \frac{e^{jwM_1} - e^{jw(M_2+1)}}{1 - e^{-jw}}$$

$$= \frac{1}{M_1 + M_2 + 1} \frac{e^{jw(M_1+M_2+1)/2} - e^{-jw(M_1+M_2+1)/2}}{1 - e^{-jw}} e^{-jw(M_2-M_1+1)/2}$$

$$= \frac{1}{M_1 + M_2 + 1} \frac{e^{jw(M_1+M_2+1)/2} - e^{-jw(M_1+M_2+1)/2}}{e^{jw/2} - e^{-jw/2}} e^{-jw(M_2-M_1)/2}$$

$$= \frac{1}{M_1 + M_2 + 1} \frac{\sin\left(\ (M_1 + M_2 + 1)/2\right)}{\sin\left(\ /2\right)} e^{-jw(M_2-M_1)/2}.$$

Frequencies around $\omega = 2\pi$ are indistinguishable from frequencies around $\omega = 0$. In effect, however, the frequency response passes only low frequencies and rejects high frequencies. Since the frequency response is completely specified by its behavior over the interval $-\pi < \omega \leq \pi$, the ideal low pass filter frequency response is more typically shown only in the interval $-\pi < \omega \leq \pi$. It is understood that the frequency response repeats periodically with period 2π outside the plotted interval.

The value of $H(e^{jw})$ represent the periodicity, which is the requirement for a discrete-time system to have the frequency response. At high frequencies, the value of $|H(e^{jw})|$ falls off. So, the high frequency extenuates and suggest that the input sequence of the system will smooth out the rapid variation. It can be assumed that the system is representing the rough approximation for a low pass filter.

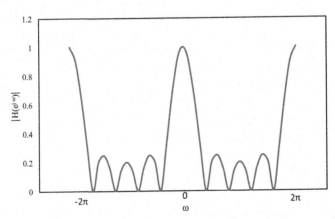

Figure 6.1   The moving-average system showing magnitude of the frequency response for the case $X_1 = 0$ and $X_2 = 4$.

Table 6.1 Typical values of magnitude of the frequency response for $\in [-2\pi, 2\pi]$ with $n = 54$

Number of Terms	w	$\left|H\left(e^{jw}\right)\right|$
1	−6.2832	1
2	−5.969	0.904013939
3	−5.6549	0.647246684
4	−5.3407	0.311499158
5	−5.0265	4.10389E-05
6	−4.7124	0.199995592
7	−4.3982	0.247210926
8	−4.0841	0.158733819
9	−3.7699	5.87993E-06
10	−3.4558	0.143167524

Absolute summability is a sufficient condition for the existence of a Fourier transform representation, and it also guarantees uniform convergence. The impulse responses are absolutely summable, since they are finite in length. Clearly, any finite-length sequence is absolutely summable and thus will have a Fourier transform representation. In the context of LTI systems, any FIR system will be stable and therefore will have a finite, continuous frequency response. However, when a sequence has infinite length, more focus should be the convergence of the infinite sum.

6.6.2 Stability of the Frequency Response of the Oscillating Impulse System

Let there exist a sequence with impulse response

$$\{I_n\} = \{1, -1, +1, -1, +1, -1, +1, -1, +1, -1, +1 \ldots\} = \{(-1)^n : \ 0 \le n \le 86\}.$$

The representation of frequency response (FR) of system is

$$H(e^{jw}) = \sum_{n=-\infty}^{\infty} I_n e^{-jwn}.$$

The frequency response function (FRF) exists if the condition of stability is satisfied by the system because the condition of stability is sufficient for FRF to exist.

$$\left|H\left(e^{jw}\right)\right| = \left|\sum_{n=-\infty}^{\infty} I_n e^{-jwn}\right|$$

$$\leq \sum_{n=-\infty}^{\infty} \left| I_n e^{-jwn} \right|$$

$$\leq \sum_{n=-\infty}^{\infty} \left| I_n \right|.$$

That is, the transient response is bounded by the sum of the absolute values of all the impulse response samples. If the right-hand side is bounded, i.e., if

$$\sum_{n=-\infty}^{\infty} \left| I_n \right| < \infty,$$

then the system is stable. It follows that, for stable systems, the transient response must become increasingly smaller as $n \to \infty$. Thus, a sufficient condition for the transient response to decay asymptotically (die out) is that the system be stable.

Thus, if I_n satisfy the condition of absolutely summable, then $H(e^{jw})$ exists. In this condition, the infinite series will uniformly converge to a function of w, which is continuous. A stable sequence is absolutely summable and all stable sequences have Fourier transforms. Hence, in this way stable system, having an absolutely summable impulse response, can made finite and contains continuous frequency response.

$$H(e^{jw}) = \frac{\cos(87w/2)}{\cos(w/2)} e^{-43jw}.$$

Figure 6.2 represent the plot of magnitude of $H(e^{jw})$ for $n = 86$ and the typical values of $H(e^{jw})$ up to first ten terms have been reported in Table 6.2. This can be expressed as

$$H(e^{jw}) = \sum_{n=0}^{8} (-1)^n e^{-jwn}$$

$$= \sum_{n=0}^{8} \left(-e^{-jw} \right)^n = 1 - e^{-jw} + e^{-2jw} - e^{-3jw} + \ldots + e^{-8jw}$$

$$= \frac{1 - \left(-e^{-jw} \right)^{87}}{1 - \left(-e^{-jw} \right)} = \frac{1 + e^{-87jw}}{1 + e^{-jw}}$$

$$= \frac{e^{87jw/2} + e^{-87jw/2}}{e^{jw/2} + e^{-jw/2}} \frac{e^{-87jw/2}}{e^{-jw/2}} = \frac{\cos(87w/2)}{\cos(w/2)} e^{-8jw/2}.$$

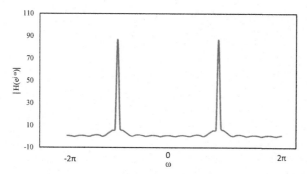

Figure 6.2 The system showing magnitude of the frequency response for the case $n = 86$.

Table 6.2 Typical values of magnitude of the frequency response for $\in [-2\pi, 2\pi]$ with $n = 86$

Number of Terms	w	$\left\| H\left(e^{jw}\right)\right\|$
1	−6.28319	1
2	−6.12609	0.854812693
3	−5.96899	0.4580518
4	−5.81189	0.083413884
5	−5.65479	0.621052955
6	−5.49769	1.001846235
7	−5.34059	1.107594056
8	−5.18349	0.887122921
9	−5.02639	0.373646363
10	−4.86929	0.31721244

Now

$$\left|H(e^{jw})\right| = \frac{\cos\left(87w/2\right)}{\cos\left(w/2\right)}.$$

Frequencies around $w = 2\pi$ are indistinguishable from frequencies around $w = 0$. In effect, however, the frequency response passes only low frequencies and rejects high frequencies. Since the frequency response is completely specified by its behaviour over the interval $-\pi < w \le \pi$, the ideal low pass filter frequency response is more typically shown only in the interval $-\pi < w \le \pi$. It is understood that the frequency response repeats periodically with period 2π outside the plotted interval.

The value of $H(e^{jw})$ represent the periodicity, which is the requirement for a discrete-time system to have the frequency response. At high frequencies, the value of $\left|H(e^{jw})\right|$ falls off. So, the high frequency extenuates and suggest that the input sequence of the system will smooth out the rapid variation. It can be assumed that the system is representing the rough approximation to a low pass filter.

Absolute summability is a sufficient condition for the existence of a Fourier transform representation, and it also guarantees uniform convergence. The impulse responses are absolutely summable, since they are finite in length. Clearly, any finite-length sequence is absolutely summable and thus will have a Fourier transform representation. In the context of LTI systems, any FIR system will be stable and therefore will have a finite, continuous frequency response. However, when a sequence has infinite length, more focus should be on the convergence of the infinite sum.

6.6.3 Stability of the Frequency Response of the Exponential System

The stability of the system can be achieved using absolute summability. Let there exist a sequence

$$\{T_n\} = \left\{1, e^{-j}, e^{-2j}, e^{-3j}, \ldots\right\}$$
$$= \left\{e^{-nj} : 0 \leq n \leq \infty\right\}.$$

The representation of frequency response (FR) of system is

$$H(e^{jw}) = \sum_{n=-\infty}^{\infty} T_n e^{-jwn}.$$

The frequency response function (FRF) exists if the condition of stability is satisfied by the system because the condition of stability is sufficient for FRF to exist.

$$\left|H(e^{jw})\right| = \left|\sum_{n=-\infty}^{\infty} T_n e^{-jwn}\right|$$
$$\leq \sum_{n=-\infty}^{\infty} \left|T_n e^{-jwn}\right|$$
$$\leq \sum_{n=-\infty}^{\infty} |T_n|.$$

That is, the transient response is bounded by the sum of the absolute values of all the impulse response samples. If the right-hand side is bounded, i.e., if

$$\sum_{n=-\infty}^{\infty} |T_n| < \infty,$$

then the system is stable. It follows that, for stable systems, for stable systems, the transient response must become increasingly smaller as $n \to \infty$. Thus, a sufficient condition for the transient response to decay asymptotically (die out) is that the system be stable.

Thus, if T_n satisfy the condition of absolutely summable, then $H(e^{jw})$ exists. In this condition, the infinite series will uniformly converge to a function of w, which is continuous. As a stable sequence is, absolutely summable and all stable sequences have Fourier transforms. Hence, in this way stable system, having an absolutely summable impulse response, can made finite and contains continuous frequency response.

6.6.3.1 Stability of frequency response of the exponential system up to *n* = 0, 1, 2,..., 88

Let there exist a sequence

$$\{T_n\} = \left\{ 1, e^{-j}, e^{-2j}, e^{-3j}, \dots, e^{-88j} \right\}$$
$$= \left\{ e^{-nj} : 0 \le n \le 88 \right\}.$$

The representation of frequency response (FR) of system is

$$H(e^{jw}) = \sum_{n=-\infty}^{\infty} T_n e^{-jwn}.$$

The frequency response function (FRF) exists if the condition of stability is satisfied by the system because the condition of stability is sufficient for FRF to exist.

Figure 6.3 represent the plot of magnitude of $H\left(e^{jw}\right)$ for $n = 88$. This can be expressed as

$$H(e^{jw}) = \sum_{n=0}^{88} e^{-jn} e^{-jwn}$$
$$= \sum_{n=0}^{88} e^{-j(1+w)n}$$
$$= 1 + e^{-j(1+w)} + e^{-2j(1+w)} + e^{-3j(1+w)} + \dots + e^{-88j(1+w)}$$

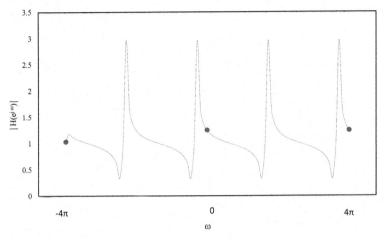

Figure 6.3   The system showing magnitude of the frequency response for the case $n = 88$.

For the geometric series with common ratio $r = e^{-j(1+w)}$

$$H(e^{jw}) = \frac{1 - e^{-89j(1+w)}}{1 - e^{-j(1+w)}}$$

$$= \frac{e^{89j(1+w)/2} - e^{-89j(1+w)/2}}{e^{j(1+w)/2} - e^{-j(1+w)/2}} \frac{e^{-89j(1+w)/2}}{e^{-j(1+w)/2}}$$

$$= \frac{\sin\left(89(1+w)/2\right)}{\sin\left((1+w)/2\right)} e^{-88j(1+w)/2}$$

Magnitude of $H(e^{jw})$

$$\left| H(e^{jw}) \right| = \left| \frac{\sin\left(89(1+w)/2\right)}{\sin\left((1+w)/2\right)} e^{-88j(1+w)/2} \right|$$

$$= \frac{\sin\left(89(1+w)/2\right)}{\sin\left((1+w)/2\right)}.$$

Frequencies around $w = 2\pi$ are indistinguishable from frequencies around $w = 0$. In effect, however, the frequency response passes only low frequencies and rejects high frequencies. Since the frequency response is completely specified by its behaviour over the interval $-\pi < w \leq \pi$, the ideal low pass filter frequency response is more typically shown only in the interval $-\pi < w \leq \pi$. It is understood that the frequency response repeats periodically with period 2π outside the plotted interval.

Table 6.3 Magnitude of the frequency response for $\in [-4\pi, 4\pi]$ with $n = 88$

Number of Terms	w	$\left|H\left(e^{jw}\right)\right|$
1	−6.28319	1.032246465
2	−6.12609	1.178464926
3	−5.96899	1.131083616
4	−5.81189	1.095609836
5	−5.65479	1.0670272
6	−5.49769	1.042603617
7	−5.34059	1.020664376
8	−5.18349	1.000054267
9	−5.02639	0.979862159
10	−4.86929	0.959248872

The value of $H(e^{jw})$ represent the periodicity, which is the requirement for a discrete-time system to have the frequency response. At high frequencies, the value of $\left|H(e^{jw})\right|$ falls off. So, the high frequency extenuates and suggest that the input sequence of the system will smooth out the rapid variation. We can also say that the system is representing the rough approximation to a low pass filter.

Absolute summability is a sufficient condition for the existence of a Fourier transform representation, and it also guarantees uniform convergence. The impulse responses are absolutely summable, since they are finite in length. Clearly, any finite-length sequence is absolutely summable and thus will have a Fourier transform representation. In the context of LTI systems, any FIR system will be stable and therefore will have a finite, continuous frequency response. However, when a sequence has infinite length, we must be concerned about convergence of the infinite sum.

6.6.3.2 Stability of frequency response of the exponential system up to n = 0, 1, 2, …, 176

Let there exist a sequence

$$\{T_n\} = \left\{1, e^{-j}, e^{-2j}, e^{-3j}, \ldots, e^{-17\,j}\right\}$$
$$= \left\{e^{-nj} : 0 \leq n \leq 176\right\}.$$

The representation of frequency response (FR) of system is

$$H(e^{jw}) = \sum_{n=-\infty}^{\infty} T_n e^{-jwn}.$$

The frequency response function (FRF) exists if the condition of stability is satisfied by the system because the condition of stability is sufficient for FRF to exist.

Figure 6.4 represent the plot of magnitude of $H(e^{jw})$ for $n = 88$. This can be expressed as

$$H(e^{jw}) = \sum_{n=0}^{17} e^{-jn} e^{-jwn}$$

$$= \sum_{n=0}^{17} e^{-j(1+w)n}$$

$$= 1 + e^{-j(1+w)} + e^{-2j(1+w)} + e^{-3j(1+w)} + \cdots + e^{-17\, j(1+w)}$$

For the geometric series with common ratio $r = e^{-j(1+w)}$

$$H(e^{jw}) = \frac{\sin\left(177(1+w)/2\right)}{\sin\left((1+w)/2\right)} e^{-17\, j(1+w)/2}$$

Magnitude of $H(e^{jw})$

$$\left|H(e^{jw})\right| = \frac{\sin\left(177(1+w)/2\right)}{\sin\left((1+w)/2\right)}.$$

Frequencies around $w = 2\pi$ are indistinguishable from frequencies around $w = 0$. In effect, however, the frequency response passes only low frequencies and rejects high frequencies. Since the frequency response is completely specified by its behaviour over the interval $-\pi < w \leq \pi$, the ideal low pass filter frequency response is more typically shown only in the interval $-\pi < w \leq \pi$. It is understood that the frequency response repeats periodically with period 2π outside the plotted interval.

The value of $H(e^{jw})$ represent the periodicity, which is the requirement for a discrete-time system to have the frequency response. At high frequencies, the value of $\left|H(e^{jw})\right|$ falls off. So, the high frequency extenuates and suggest that the input sequence of the system will smooth out the rapid variation. We can also say that the system is representing the rough approximation to a low pass filter.

Absolute summability is a sufficient condition for the existence of a Fourier transform representation, and it also guarantees uniform convergence.

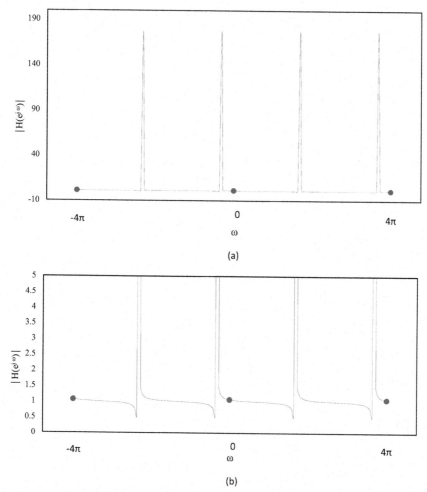

Figure 6.4 The system showing magnitude of the frequency response for the case $n = 176$.

The impulse responses are absolutely summable, since they are finite in length. Clearly, any finite-length sequence is absolutely summable and thus will have a Fourier transform representation. In the context of LTI systems, any FIR system will be stable and therefore will have a finite, continuous frequency response. However, when a sequence has infinite length, we must be concerned about convergence of the infinite sum.

Table 6.4   Magnitude of the frequency response for $\in [-4\pi, 4\pi]$ with $n = 176$

Number of Terms	w	$\left|H\left(e^{jw}\right)\right|$
1	−6.28319	1.064169
2	−6.12609	1.05443
3	−5.96899	1.04664
4	−5.81189	1.040209
5	−5.65479	1.034758
6	−5.49769	1.030034
7	−5.34059	1.025862
8	−5.18349	1.022113
9	−5.02639	1.018694
10	−4.86929	1.015531

6.7 Conclusions

The theory of summability has many uses throughout analysis and applied mathematics, for example, in analytic continuation, quantum mechanics, probability theory, Fourier analysis, approximation theory and fixed-point theory. The methods of almost absolute summability and statistical summability have become an active area of research in the recent years. They are used in applied analysis for the generation of iterative methods in solution of the linear system of equations and for both the creation and acceleration of convergence of a Fourier series in approximation theory. The Fourier series of functions in the differential equation give some prediction about the dynamics of the solution of differential equation.

The summability techniques are wildly used in approximating the signals of digital filters such as finite impulse response filters [FIR filters] and infinite impulse response filters [IIR filters]. This feature of summability has a variety of applications like in signal processing, audio signal processing, speech and image processing, communications, radar, sonar, medical signal processing, etc. For an engineering problem having piecewise finitely differentiable data with jump discontinuities, there are more sophisticated ways to use the sampled data to obtain a more accurate solution using the approximations of the Fourier coefficients by summability techniques. With the help of a minimal set of sufficient conditions for a model, the error can be minimized by using summability methods so that the output data is filtered and according to our interests.

References

[1] N. H. Abel, Üter suchungen uöber die reihe $1 + mx + \frac{m(m-1)}{2!}x^2 +$ ... *Journal für die reine und angewadte Mathematik, crelles*, (1826) 311–339.

[2] A. L. Cauchy, Cours d'analyse de lécole polytechinque Part I, *Analyse Algébrique*, (1821).

[3] E. Cesàro, Sur la multiplication des series, *Bull. Sci. Math.*, 14 (2) (1890).

[4] L. S. Bosanquet, Note on the absolute summability (C) of a Fourier series, *Journal of London Math. Society*, 11 (1936a) 11–16.

[5] L. S. Bosanquet, The absolute Cesàro summability (C) of a Fourier series, *Proceedings of the London Math. Society*, 41(2) (1936b) 517–528.

[6] H. C. Chow, On the absolute summability (C) of power series, *Journal of the London Math. Society*, 14 (1939) 101–112.

[7] J. M. Hyslop, on the absolute summability of Fourier series, *Proc. London Math. Soc.*, 43 (2) (1937) 475–483.

[8] F. T. Wang, Note on the absolute summability of Fourier series, *Journal London Math. Soc.*, 16 (1941) 174–176.

[9] H. Bor, On the absolute φ-summability factors of infinite series, *Portugaliae mathematica*, 45 (2) (1988) 131–138.

[10] H. Bor, A note on absolute Nörlund summability, *J. Inequal. Pure Appl. Math,* 6 (62) (2005) 1–5.

[11] H. S. Özarslan, A note on absolute summability factors, *Proc. Indian Acad. Sci.-Math. Sci.*, 113 (2) (2003) 165–169.

[12] H. S. Özarslan, T. Ari, Absolute matrix summability methods, *Appl. Math. Lett.*, 24 (12) (2011) 2102–2106.

[13] W. T. Sulaiman, Applications of generalized power increasing sequences, *Pacific Journal of Applied Mathematics*, 3 (4) (2011) 233–240.

[14] W. T. Sulaiman, On some generalization of absolute Cesàro summability factors, *J. Classical Analysis*, 1 (1) (2012) 23–30.

[15] A. V. Oppenheim and R. W. Schafer, Discrete-Time Signal Processing, third edition, 1989.

7

Design of Manufacturing, Control, and Automation Systems

Mohit Pant[1,*], Mohit Sood[2], Aeshwarya Dixit[3] and Sahil Garg[1]

[1]Department of Mechanical Engineering, National Institute of Technology, Hamirpur, Himachal Pardesh-177001, India
[2]Department of Mechanical Engineering, Rayat and Bahara Institute of Engineering and Nano Technology, Hoshiarpur, Punjab-146104, India
[3]University School of Business, Chandigarh University, Gharuan, Punjab-140301, India
E-mail: mohitpant.iitr@gmail.com; msood04@gmail.com; aeshwaryagarg1017@gmail.com; sahil.garg1017@gmail.com
*Corresponding Author

This work is related to various parameters that are considered in the design of the manufacturing control and the automation. Manufacturing system can be designed in accordance to the techniques used to provide final state of product that may be finished or semi-finished. Additive subtractive and rapid prototypes are being used in the manufacturing design. The design of control system is done by two types one in the open type control system and other is the closed type control system. Feedback loop is absent in the open type control system. Modern control system is designed with computer system that can be changed according to computer programs. Automation can be designed according to type of flow and other volume of product that must be controlled.

7.1 Introduction

Before you start reading this chapter, pause and scrutinize your surroundings. There are many objects around you like chair, cell phone computer, light fixture and mechanical pencil. While observing them you will realize that

159

all these objects, including various individual components, are made from a wide variety of materials which have been assembled and produced into various useful objects. There are many objects which are soft and are made from one component like paper clip, nail, spoon, and door key. Whereas many other objects consist of numerous individual pieces that are built and assembled by a combination of processes called manufacturing. The word **"manufacture"** is derived from two Latin words **"menu** and **facts"**, which mean "made by hand" and appeared in English in 1567. In 1683, the word **"manufacturing"** first immerged, whereas in 15th century **"production"**, came nonexistence which is many times used interchangeably with the word manufacturing. Manufacturing is concerned with making products [1]. But many times, manufactured products may be used to produce other products, such as

(a) Press, to shape flat sheet metal into automobile bodies,
(b) Drill, for producing holes,
(c) Industrial sewing machines, for making clothing at high rates, and
(d) Numerous pieces of machinery,

These small or large individual items ranging from Pen nib to electric motors to crankshafts and connecting rods for automotive engines are used for manufacturing different products wherein they are a part of final product. It is also noteworthy that these small objects like needle, nuts, paper clips, safety pin are also individual products which are being manufactured individually. Contrary to this, there are other products which used directly as end product like plastic pipes, spool of wire. Products like these are used directly by cutting them as per the requirement [2]. As the process of manufacturing typically starts with procurement of raw material ,which is then processed to make finished goods , the raw material also has certain monetary value even before processing. For example, iron has monetary value even before it is processed into knives, nails, cutlery, pans, plumbing fixtures etc. After the manufacturing process value is added to the raw material. Similarly, final products have more value as compared to the raw material used to make it. Products such as play station, air conditions, Athletic shoes are thus called value added products.

Historically, the commercial production of goods started with manufacturing of household items which were made from wood, stone or metal. Initially, materials like gold copper and iron were used to make utensils and decorative items thereafter silver, led, tin, bronze and brass were also used for same purpose. In the beginning, casting was used as the processing method as it is easier process. But with changing times and technological innovation, this simple molding method developed into more complex process to facilitate faster, cheaper, and quality industrial production for instance, during the 16th and 17th centuries lathe machines were chief sources of screw thread manufacturing processes, but it was not until some three centuries later that automatic screw machines were developed.

The change in traditional manufacturing methods, through manual labor, occurred after first industrial revolution in England which began in the 1750s. In the mid-1900s, second industrial revolution began with improvement of solid-state electronic devices and computers [3]. England and Europe were pioneers in this change in methods of manufacturing which occurred initially in textile industries and manufacturing industries involving metal cutting procedures. Further development in manufacturing began in 1800 century as an American Eli Whitney improvised the use of machinery with replaceable part, improved design, and speeding up of production process [4]. The introduction of replaceable parts has revolutionized the manufacturing industry which was not the case earlier as it was considered almost impossible that two manufactured parts can meet similar specifications [5]. Nowadays even a smallest nut or pin can be made with specific accuracy and can be replaced on a machine even after decades of use. The early 1940s was a golden period for manufacturing industry with innovation in numerous household and industrial goods. These innovative products were new to the world and consumer market. Many benchmarks were achieved during this era of research and innovation in technological advancement and manufacturing. While studying the history of manufacturing, the gradual growth can be traced back to Roman era ($\sim$500 B.C. to 476 A.D.) when glassware, cutlery, weapons were all made manually involving tedious and tiresome labor in factories of Rome. In today's high-tech manufacturing process ,the tiresome labor is replaced with time saving and fast mass production like (a) aluminum beverage cans are made at rates of more than 500 per minute, with each can costing about four cents to make, (b) holes in sheet metal are punched at rates of 800 holes per minute, and (c) incandescent light bulbs are made at rates of more than 2000 bulbs per minute each costing less than one dollar.

7.2 Steps in Design of Manufacturing Systems

Before understanding the type of manufacturing is also necessary to understand how the design of manufacturing works. It starts form the demand in the market and definition of the product that is needed as shown in Figure 7.1.

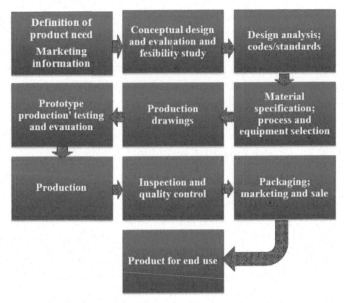

Figure 7.1   Flowchart of manufacturing.

7.3 Types of Manufacturing

The manufacturing can be divided according to the following types

1. Additive manufacturing
2. Subtractive manufacturing
3. Rapid prototyping

7.4 Additive Manufacturing (AM)

As the name suggests this process involves addition of material to complete the desired product. Most of the manufacturing processes are involved with removal of materials but this process employs layer by layer addition of material to achieve the product. "Solid freeform fabrication" (SFF) technologies is the name given to such class of processes. The technologies involving addition of materials are 3D printing, rapid prototyping, direct digital manufacturing etc. Earlier due to high cost these technologies were involved in prototype manufacturing only. But now these technologies have found their way in for bio-medical applications. Most recently a prototype of human lung was grown in International Space Station using 3D printing. Dentures have been produced using rapid prototyping also knee joints, hip bone replacement joints are nowadays made using additive manufacturing techniques owning to their accuracy. These customized products have paved the way for rapid prototyping techniques to more common use and have given a new domain from narrow rapid prototyped products [6]. Some advancements in the field of additive manufacturing processes include Vat photopolymerization, Material jetting, Binder jetting etc. A comparative study was conducted by the World Technology evaluation centers (WTEC) on integrative A/S approaches to material synthesis and manufacturing in United States and Europe in the framework of direct manufacture of products. The report not only reviews the techniques of A/S manufacturing but has given a peek into the application of these techniques for advanced bio-medical uses like tissue modeling, implants construction etc. [7].

Additive manufacturing is the process of manufacturing in which the product is formed with the addition of the layers into the from the third-party material. For example, welding, the welding is the well-known process of the additive manufacturing.

7.4.1 Design of Additive Manufacturing System

Additive Manufacturing (AM) has recently become a higher end research topic due to its ability to provide limitless applications in restructuring and remodeling of parts. Technologies like rapid prototyping are increasingly used for bio-medical applications. The National Science Foundation even organized a workshop to bridge the gap between industry and research and to motivate for the use of AM. This section of the chapter is aimed at discussing a brief of points discussed in that workshop to help advance the AM procedures in forwarding the technology:

I. Conjunction of AM and traditional manufacturing processes so that the present structure of industries can be benefited.
II. Proper selection of AM processes and material selection on the merit of material properties fit for a particular process, need and utilizzation.
III. Incorporating the use of computational tools to avoid excessive costs in AM techniques [5].

7.4.2 Subtractive Type Manufacturing

As the name suggests the process of subtractive manufacturing involves removal of material from a 3D block and reducing it to desired shape.

Earlier manual cutting of metal was employed which is now replaced by advanced CNC machining process. The tool holders as well as the job holders move along all possible axis of rotation and also in sync with each other to conceive the desired product mitigating the need of flipping the block as in manual processes. An example that established the supremacy of subtractive manufacturing is the ability to machine an extremely thin piece of plastic into a living hinge which is not yet possible for a 3D printer. For those prototypes that require living hinge components, it is useful to produce certain parts using additive manufacturing while using the CNC machine for specialty components like a living hinge. The technological juggernaut has continuously evolved the term machining over the past 1500 years and its meaning is not just limited to removal of materials. Going by the general timeline a person in the 18th century working mostly by using hand tools and performing operations involving wooden carving, forging, fitting or involved in general assembly or repair or machined was called as "machinist". With development of engines the inventors like James Watt started to fir in the def-inition of machinist, but still the noun and verb form of the word machinist i.e. machined, and machining did not yet exist. These came into existence in the 19th century and were now recognized as worldwide concepts. The dawn of machine age brought the classification of machining processes and machines. Now the processes like turning, boring, planing, milling, tapping, sawing, shaping, drilling, reaming, and broaching had their very own existence and dedicated specialists for each kind of process. The development of these pro-cesses evolved the cutting tools associated with them and continuous efforts to improve the metal removal rate and efficiency were made. In the modern age the machining processes are now classified as conventional and uncon-ventional processes. The conventional processes involve traditional processes like turning drilling etc. and the non-conventional ones new technologies such as electrical discharge machining (EDM), electrochemical machining (ECM), electron beam machining (EBM), ultrasonic machining etc.

7.4.3 Machining Process in Subtractive Manufacturing

As any machining process is performed the workpiece so obtained has a different surface finish and sometimes the surface microstructure varies as well due to grain cutting or heat treatment received due to frictional heat. The process of turning can be simply defined as removal of material from the workpiece which is experiencing constant rotation and the single point cutting tool is applied on its surface to remove material in form of chips.

The most common machine performing this operation is lathe and the advanced version in CNC machine. Materials having a range of hardness from wood to steel to Titanium can be machining via turning process. Various machining processes can be performed on a lathe similar to turning are facing, tapering, knurling etc. Next comes a conventional drilling process employed to create holes in a job. The drilling tool is a multipoint cutting tool that has two or four helical cutting edges. The direction of feed of tool and axis of rotation of the tool are in line to the workpiece. After successful generation of hole in the workpiece the hole is smoothened, or its diameter is adjusted by an operation known as boring. The tool employed for a boring process involves a single bent pointed tip is pushed into the drilled hole to improve its finish, so boring can be regarded as a finishing process. Another process performed to remove material from an already drilled hole is called reaming and it is regarded as much finer process as the material removed is very small. Another process known as milling is performed to obtain plane or straight surface with the help of a rotating tool with multiple cutting edges which is moved slowly relative to the material. Other conventional machining operations include shaping, planing, broaching and sawing. Also, grinding and similar abrasive operations are often included within the category of machining.

7.4.4 Cutting Tools for Subtractive Manufacturing

Cutting tool is employed to remove material from the workpiece with help of one or more cutting edges e.g. a single point cutting tool is used lathe whereas a milling cutter has many cutting edges. Some examples of processes involving the use of single point cutting tool are planning, boring, turning etc. When machining is performed the edge of the tool goes through the top surface of the workpiece. In case of multi point cutting tools, the cutting edge is rotated with respect to the workpiece to achieve the desired machining. There is a contrasting difference between the shapes of single point and multi point cutting tools although the tool geometries and other vital tool elements can be similar. The major property that helps in selection of the cutting tool is the hardness which should be greater than the harness of material to be cut. There are two major parts of a cutting tool:

- The rake face
- The flank

The job of the rake face is to direct the material removed in form a chip to a desired angle called as the rake angle so that smooth machining operation can be carried out. The rake angle is measured along the plane perpendicular to the work surface. The value of rake angle defines the type of machining operation performed. Positive rake is used for softer materials while negative rake is sued for harder materials and the neutral rake is avoided as it causes carter and wear. To protect the newly formed surface of the workpiece from sudden abrasion or wear flank comes into play and the angle between the workpiece and flank is called as the relief angle.

7.5 Rapid Prototyping

Rapid prototyping (RP) is a new manufacturing technique that allows for fast fabrication of computer models designed with three-dimension (3D) computer-aided design (CAD) software. RP is used in a wide variety of industries, from shoe to car manufacturers. This technique allows for fast realizations of ideas into functioning prototypes, shortening the design time, leading towards successful final products. RP technique comprises of two general types: additive and subtractive, each of which has its own pros and cons [9]. Subtractive type RP or traditional tooling manufacturing process is a technique in which material is removed from a solid piece of material until the desired design remains. Examples of this type of RP includes traditional milling, turning/lathing or drilling to more advanced versions – computer numerical control (CNC), electric discharge machining (EDM). Additive type RP is the opposite of subtractive type RP. Instead of removing material, the material is added layer upon layer to build up the desired design such as stereo lithography, fused deposition modeling (FDM), and 3D printing.

7.5.1 Advantages and Disadvantages of Rapid Prototyping

Subtractive type RP is typically limited to simple geometries due to the tooling process where the material is removed. This type of RP also usually takes a longer time, but the main advantage is that the end product is fabricated in the desired material. Additive type RP, on the other hand, can fabricate most complex geometries in a shorter time and lower cost. However, additive type RP typically includes the extra post-fabrication process of cleaning, post-curing or finishing.

Here are some of the general advantages and disadvantages of rapid prototyping [1]:

Advantages:

1. Fast and inexpensive method of prototyping design ideas
2. Multiple design iterations

3. Physical validation of design
4. Reduced product development time

  Disadvantages:

1. Resolution not as fine as traditional machining (millimeter to sub-millimeter resolution)
2. Surface flatness is rough (dependent of material and type of RP)

7.5.2 Types of Rapid Prototypes

➢ Additive Rapid Prototyping
➢ Liquid base – Stereo Lithography Apparatus (SLA)
➢ Inkjet-based printing
➢ Solid based: Fused Deposition Modeling (FDM)
➢ Powder Base: Selective Laser Sintering (SLS)

Additive Rapid Prototyping: The different types of additive RP technologies can be categorized into three types: liquid-based (SLA and Inkjet-based Printing), solid-based (FDM), and powder based (SLS). These are just a few examples of the different RP technologies in existence. Regardless of the different types of RP technologies, all of them require the 3D CAD model's STL file for fabrication. These STL files are then used to generate to 2D slice layers for fabrication.

Liquid base – Stereo Lithography Apparatus (SLA): SLA RP technology has three main parts: a vat filled with ultraviolet (UV) curable photopolymer, a perforated build tray, and a UV laser, (Figure 7.2). Figure 7.3 shows a production level SLA system by 3D SYSTEMS. The fabrication process starts with positioning the build tray a slice layer depth below the surface of the photopolymer. A slice layer is cured onto the build tray with the UV laser. The pattern of the slice layer is "painted" with the UV laser with the control of the scanner system [5]. Once the layer is cured, the tray lowers by a slice layer thickness allowing for uncured photopolymer to cover the previously cured slice. The next slice layer is then formed on top of the previous layer with the UV laser, bonding it to the previous layer. This process is repeated until the entire 3D object is fully formed. The finished 3D object is removed and washed with solvent to remove excess resin from the object. Finally, the object is placed in a UV oven for further curing. During the fabrication process, support scaffolding can be fabricated to support overhangs or undercuts of the 3D object. These can be cut off after fabrication [8].

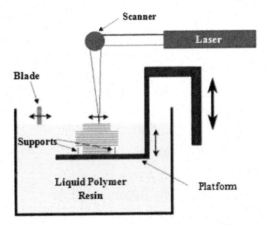

Figure 7.2 Stereo Lithography Apparatus (SLA).

Source: http://www.additive3d.com

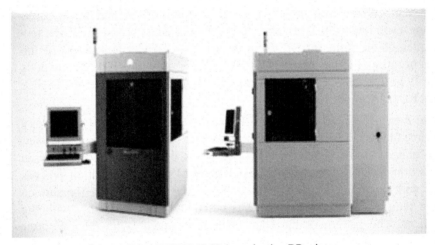

Figure 7.3 SYSTEMS SLA production RP printers.

Source: http://www.3dsystems.com

Inkjet-based Printing: This RP technology is similar to the SLA technology, both of which utilize UV curable photopolymer as the build material. Two types of UV curable photopolymer materials are used: a model that acts as the structure and support material acting as scaffolding to the object. The technology is based on inkjet systems as shown in Figure 7.4 where it has 'ink' cartridges and a print head. During fabrication, a thin layer

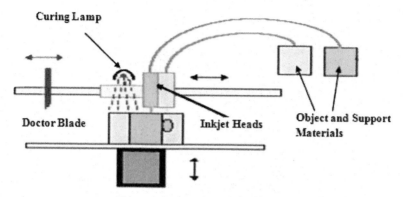

Figure 7.4 Inkjet-based printing.

Figure 7.5 Stratasys inkjet-based RP printer.

Source: http://www.stratasys.com

of photopolymer is jetted on to the build tray. The jetted photopolymer is cured by UV lamps mounted to the side of the print heads. Next, the tray lowers by precisely one layer's thickness, allowing for the next slice to be jetted on to the previous slice [10]. This process repeats until the 3D object is formed. Once completed, the support material is removed with a high-pressure washer. A commercially available inkjet-based RP printer by Stratasys is shown in Figure 7.5.

***Solid based: Fused Deposition Modeling (FDM)*:** FDM RP technologies use a thermoplastic filament, which is heated to its melting point and then extruded, layer by layer, to create a three-dimensional object, shown in Figure 7.6. Two kinds of materials are used: a model material which acts

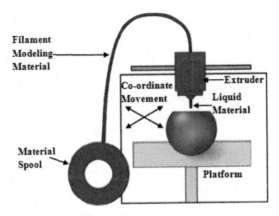

Figure 7.6   Fused deposition modeling.

Source: http://solidfill.com/en/Fused_Deposition_Modeling

Figure 7.7   3D SYSTEMS FDM desktop RP printer.

Source: http://www.3dsystems.com

as the structure and a support material which acts as a scaffolding to support the object during the fabrication process During the fabrication process, the filaments are fed to an extrusion nozzle unwounded from a coil. This nozzle is heated to melt the filament which is then extruded on to a build tray forming a slice of the 3D object as cools and hardens. Next, the build tray is lowered, or the extrusion nozzle is raised, by a thickness of an extruded layer, for the next slice layer to be extruded on top of the previous layer. As the extruded thermoplastic cools, it also binds to the previous layer. This continues until all the slices are printed to finally form the full 3D object. After the fabrication process, the support build material is typically dissolved by water if the water-soluble wax was used or broken off if polyphenylsulfone was used. An affordable desktop version by 3DSYSTEM is shown in Figure 7.7.

7.6 Control

In engineering world, the term control may refer to as a set of algorithms which are tasked to perform a certain operation and generate feedback from the results of operation performed. The simplest example of control can be attributed to traffic signal control [11].

Modern engineering or tech giants are working on high end technologies like drones, autonomous vehicles, artificial intelligence (AI)-based real-time management systems and CYBORGs i.e. biologically engineered systems [12]. The core of these systems can be called as core control which processes the information generated from the output of the function of the machines and stores it also analyses it. The AI based control systems interprets the output of an operation and compares is with set of desired parameters to update it or modifiy it to be better.

Now the term control can be summed up as this loop of observing the action performed, sensing the output, computation of the results and modification in action to evolve to be better. A control logic is a complex network of codes designed to generate desired results and feedback and it is very important that the system dynamics of the loop are stable [13]. The system is tested or rather simulated in the presence of all plausible uncertainties using various simulation procedures like Monte Carlo simulation tests or simulations based on interchangeable variable of a system to trap the essential responses of the system. This helps is prediction of possible future failures or breakdowns resulting in loss of efficiency and productivity. A characteristic case of a recent type of control system is shown in Figure 7.8, the illustration clarifies the elementary components of sensing, computation and actuation. The computer system nowadays work on a digital platform requiring the conversion of analog information into digital form and vice-versa using specific equipment called as converters [14]. Control laws in form of algorithms are defined to operate the system and extract the desired outputs. An external operator is used to provide input commands to the system to influence or control the effective working of system. Control engineering is a blend of sciences e.g. Physics defines the laws of dynamics, Computer science defines

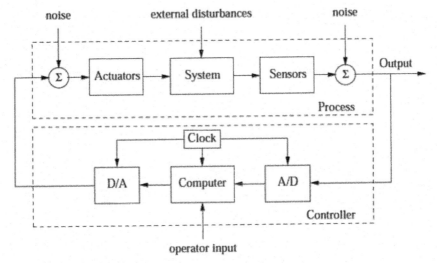

Figure 7.8   Components of a computer-controlled system.

the codes or algorithms, operations research is used to optimize and check the probabilities in the system [15]. The use of modeling techniques in control helps in establishing high-end interconnection between the elements of the system which is crucial for large scale engineering structures [16]. The development of control systems has given a big boost to the computer science engineering and language-based programming software as most of the algorithms are designed it. Although the implementation of algorithm and designed code for the machine work very differently taking into consideration all possible alterations of the machine. Moreover, the control system also incorporates preventive measures to avoid the errors and breakdown possibilities.

7.6.1 Feedback in Control System

When a system is designed in such a way that the output generated by two or more dynamic entities of a system interdependent or correlated to each other such that the closed loop completing the cause and effect algorithm and outputs of the system are re-passed as inputs defining proper working of the system is called as feedback. A dynamic system is designed to intelligently change its behavior with respect to time corresponding to the external effects like change is physical quantities like temperature, force etc. Feedback can

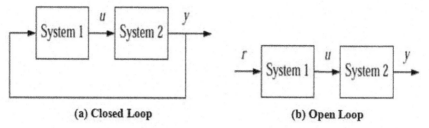

(a) Closed Loop   (b) Open Loop

Figure 7.9 Open and closed loop systems. (a) The output of system 1 is used as the input of system 2 and the output of system 2 becomes the input of system 1, creating a "closed loop" system. (b) The interconnection between system 2 and system 1 is removed and the system is said to be "open loop".

also be seen as a looped argument in which the dynamics of subsystems affect each other in a to and fro manner thus, making a casual reasoning difficult [17]. Henceforth, it is imperative the analyze the system as a whole rather than introspecting smaller parts of system. This counter-intuitive behaviour of systems has motivated the use of formal methods to comprehend the systems. Figure 7.9 exemplifies block diagram layout of the idea of feedback system. A generalized form of feedback system is categorized into open loop and closed loop system. A simple understanding of a closed loop system is that if the actions and reactions or inputs and outputs of the system are interlinked to form a cycle is referred to as closed loop (Figure 7.9(a)). On the contrary the open loop system does not complete the cycle of events or the interconnection is broken (Figure 7.9(b)). Biological systems [18] are excellent examples of feedback systems interconnecting even the tiniest elements of human anatomy. All the action commands generated by brain the fed back to the brain the form of results of actions.

7.7 Automation

With the dawn of globalization and neck to neck competition between industries and global economies the demand of finished products manufacturing industries shifted from manual methods to automation tools. The automatic machinery not only completes the product but also has capability to finish it and pack it in desired numbers which has eliminated bottleneck human errors and reduced considerable amount of time. Moreover, automation has reduced errors to almost zero by performing precise operations without any lag.

7.7.1 Application of Automation in Industries

The goal of every manufacturing unit is to make profit and it is possible if the input cost is minimized as compared to the market cost or selling price of the product. Many variables define input cost of the product in an industry e.g. cost of material, choice of manufacturing process, labor cost etc. Automation has minimized the labor cost and time delays. Automation has a huge impact on all the input parameters defining the input cost [19]. For instance, production time is greatly affected by machine breakdown or changeover time for tools or workpiece. If there is any change in the design or configuration of the product the whole production line suffers and causes delay. This causes the expensive machinery to stay idle and become unproductive [20]. Automatic CNC controlled machines have inbuilt changeover input adjustments with various tools changers and centers inbuilt. The design configuration of any configuration whether conventional or modified can be fed into the machine in the form of co-ordinates and the product can be finalized without any delay or break in the production process [21].

7.7.2 Type of Automation System

Fixed Automation: The most widely used automation process in large scale industries developing similar kind of parts for example automobile units, paint shops, distillations plants [22] etc. employ fixed automation systems. The process of fixed automation involves fabrication of a defined design of product to be manufactured in a continuous pattern. The design input parameters are fed to the machine and these machines are dedicated in manufacturing high production volumes of the product. The machinery involved in this automation process is under repetitive process and hardly any change is made for a long duration of time. The system fails only when the product design has become obsolete.

Programmable Automation: If an industry is involved in production of batches of various products programmable automation does the trick for them. For instance, a company having programmable automation unit has received an order of two kind of products involving machinery of same kind. programmable automation units work in a way that first the engineers can input the variables required for a product and finish the required amount of that product and then they can change the input parameters on the same machines to manufacture the different product. However, non-trivial programming effort may be needed to reprogram the machine or sequence of operations [23]. Apart from initial investment the programmable automation unit provides good return to the company and some examples of industries employing this process are Steel Rolling Mills, Paper Mills etc.

Flexible Automation: The evolution of programmable automation gave rise to flexible automation. The drawback with programmable automation is that the whole system of automation suffers if the changes are made in the computer code operating the machines. Flexible automation allows the engineer to give inputs to the products on any stage of the product fabrication such that the former process involved in the manufacturing process either automatically get adjusted or are unperturbed [24]. Flexible automation is known to reduce health and safety hazards along with reducing the fixed production costs. The product processes involved in the fabrication of the product are changes automatically and the product under fabrication is transferred from one station to the other without any human intervention thereby eliminating the time delay [25]. Nowadays robots are considered as a huge part of flexible automation process and industries employing these systems have low or medium production output volumes.

Integrated Automation: A fully automated plant involving minimal human intervention and comprising of interconnected, coordinated operations involving all machine operations and digital information processing. The integrated automation is like a network of small manufacturing units working from extraction of raw material to finishing the product and even packing it if required. The process scheduling is this automation system is very precise and all the machines under this process are controlled by a central computer system. Some advanced units are capable of analyzing the business models so that the rate of production can be estimated. Typical examples of such technologies are seen in Advanced Process Automation Systems and Computer Integrated Manufacturing (CIM).

7.8 Conclusion

Additive, subtractive and rapid prototyping is the technique designed for the manufacturing which can be used with to produce various types of products according to their complexity and requirement of the production procedure. Additive and subtractive both related to material addition and subtraction but rapid prototyping is related to printing with three-dimensional printers. Computerized control system is design according to requirement of feedback from the system. Automatic controlled system is designed with closed feedback loop. Fixed Automation used in high volume production with dedicated equipment, which has a fixed set of operation and designed to be efficient for this set. Programmable Automation is used for a changeable sequence of operation and configuration of the machines using electronic controls. Flexible Manufacturing Systems is a computer controlled. Integrated Automation is automation of a manufacturing plant, with all processes functioning under computer control and under coordination through digital information processing.

References

[1] J. E. Harwood and D. J. Huyser, "Automated kjeldahl analyses of nitrogenous materials in aqueous solutions," *Water Res.*, vol. 4, no. 8, pp. 539–545, 1970.

[2] S. Ambriz et al., "Material handling and registration for an additive manufacturing-based hybrid system," *J. Manuf. Syst.*, vol. 45, pp. 17–27, 2017.

[3] D. A. Arcos-Novillo and D. Güemes-Castorena, "Development of an additive manufacturing technology scenario for opportunity identification—The case of Mexico," *Futures*, vol. 90, pp. 1–15, 2017.

[4] M. Attaran, "The rise of 3-D printing: The advantages of additive manufacturing over traditional manufacturing," *Bus. Horiz.*, vol. 60, no. 5, pp. 677–688, 2017.

[5] C. Cozmei and F. Caloian, "Additive Manufacturing Flickering at the Beginning of Existence," *Procedia Econ. Financ.*, vol. 3, no. March 2009, pp. 457–462, 2012.

[6] S. Afazov et al., "A methodology for precision additive manufacturing through compensation," *Precis. Eng.*, 2017.

[7] D. R. Eyers and A. T. Potter, "Industrial Additive Manufacturing: A manufacturing systems perspective," *Comput. Ind.*, vol. 92–93, pp. 208–218, 2017.

[8] B. H. Jared et al., "Additive manufacturing: Toward holistic design," *Scr. Mater.*, vol. 135, pp. 141–147, 2017.

[9] P. Parandoush and D. Lin, "A review on additive manufacturing of polymer-fiber composites," *Compos. Struct.*, vol. 182, no. August, pp. 36–53, 2017.

[10] J. P. Rudolph and C. Emmelmann, "Analysis of Design Guidelines for Automated Order Acceptance in Additive Manufacturing," *Procedia CIRP*, vol. 60, pp. 187–192, 2017.

[11] L. Ramli, Z. Mohamed, A. M. Abdullahi, H. I. Jaafar, and I. M. Lazim, "Control strategies for crane systems: A comprehensive review," *Mech. Syst. Signal Process.*, vol. 95, pp. 1–23, 2017.

[12] S. A. Timoshin, "Control system with hysteresis and delay," *Syst. Control Lett.*, vol. 91, pp. 43–47, 2016.

[13] P. C. Engineer et al., "Control Systems Engineer," pp. 1–11.

[14] M. Gopal, I. J. Nagrath, and M. Gopal, "Control Systems Engineering," *New Age International Publishers*. p. 912, 2009.

[15] V. Psyk, D. Risch, B. L. Kinsey, A. E. Tekkaya, and M. Kleiner, "Electromagnetic forming - A review," *J. Mater. Process. Technol.*, vol. 211, no. 5, pp. 787–829, 2011.

[16] K. Åström and R. Murray, "Feedback systems: an introduction for scientists and engineers," no. July, p. 412, 2008.

[17] A. Chaves, M. Rice, S. Dunlap, and J. Pecarina, "Improving the cyber resilience of industrial control systems," *Int. J. Crit. Infrastruct. Prot.*, vol. 17, pp. 30–48, 2017.

[18] M. Khusairi osman, "Intelligent Control Systems," *Intell. Control Syst.*, vol. 103, no. October 2016, p. 200, 2016.

[19] U. Automation, "1 Automation Overview," *Analysis*, pp. 1–4, 2016.

[20] K. Lau and R. J. Hocken, "A Survey of Current Robot Metrology Methods," *CIRP Ann. - Manuf. Technol.*, vol. 33, no. 2, pp. 485–488, 1984.

[21] P. Kopacek, "Automation and TECIS," *IFAC-PapersOnLine*, vol. 48, no. 24, pp. 21–27, 2015.

[22] S. Pallavi, "Lesson Introduction to Industrial Automation and Control," pp. 1–18, 2013.

[23] I. Nielsen, N. A. Dung Do, Z. A. Banaszak, and M. N. Janardhanan, "Material supply scheduling in a ubiquitous manufacturing system," *Robot. Comput. Integr. Manuf.*, vol. 45, pp. 21–33, 2017.

[24] D. K. Sharma, *Overview of Industrial Process Automation.* 2011.

[25] Q. Li, I. Kucukkoc, and D. Z. Zhang, "Production planning in additive manufacturing and 3D printing," *Comput. Oper. Res.*, vol. 83, pp. 1339–1351, 2017.

8

SEIR – Application for Crop Through Water and Soil Texture

Nita H. Shah*, **Ekta N. Jayswal, Moksha H. Satia**
and Foram A. Thakkar

Department of Mathematics, Gujarat University, Ahmedabad, 380009, Gujarat, India
E-mail: nitahshah@gmail.com; jayswal.ekta1993@gmail.com; mokshasatia.05@gmail.com; foramkhakhar3@gmail.com
*Corresponding Author

Crop is a backbone of agriculture economy of any country. Soil and water are two key resources that directly or indirectly affect the crop production. The actual capacity of soil to retain the water makes that soil fertile which is necessary for ideal crop growth. To assess the effect of water and soil on the crop production, we have formulated the system of non-linear differential equations. The model is followed by its stability by finding equilibrium points; it gives the conditions which should be satisfied to maintain crop growth. The proposed model is validated through numerical simulation. The important economical results are deduced.

8.1 Introduction

8.1.1 Dynamical System

Mathematical modeling plays a significant role in controlling disease spread. One can observe the dynamic behavior of original data given in the model to see the perspective results. Of all mathematical modeling, the model using compartments is very useful technique. In this technique, the population is divided into compartments. There are many models like SIR, SIS, SEIR,

179

SEIS, etc. The SIR model is one of the simplest compartmental model and many models are derivations of this basic form. They mainly deal with epidemiology. In population science, the analysis of the distribution and determinants for health and diseases is known as epidemiology. One of the simplest models SIR consists of susceptible (S), infected (I) and removed (R) compartments. Individual remains infected for some time, though not infectious. This term is defined as exposed (E). This extension of model is SEIR model. These models are usually examined through the system of ordinary differential equations. In concluding results about stability of system means that disease spread whether it is decreased or completely dies out. The model depends on threshold value; known as basic reproduction number usually denoted by R_0. If $R_0 < 1$ then the disease-free equilibrium will be locally asymptotically stable while when, $R_0 > 1$ then the system is unstable states that the infection will spread in a population. The global stability about the equilibrium point of the model is worked out using Lyapunov function. There are several ways to find Lyapunov functions. First decide the form of Lyapunov function (e.g., quadratic), then it is parameterized by some parameters so that the required hypotheses hold. To predict the future results, numerical simulation is done using MATLAB.

8.1.2 Dynamics of Crop using Fertile Soil

Farming is one of the oldest businesses in the world and often presented as a pleasant way of life. The country produces numerous crops by farming, in area of pharmaceutics as well as cereal crops. These crops are used for various purposes for human need like for food, in industries, for animal feed etc. Farming needs land to farm. There are different types of land, either it may be fertile soil or infertile soil. Infertile soil uses for industrialism, while fertile can be used for farming. When fertile soil is crushed in the hand, it can be seen that it is composed of all kinds of particles of different sizes. There are few types of particles namely mineral, organic, etc. Mineral particles are those particles which are emerge from the degradation of rocks; where organic particles are residues of crops or animals. The soil particles seem to touch each other, but in reality, they have spaces in between. These spaces are called pores. When the soil is dry, the pores are mainly filled with air. But crops need air and water in the soil. For that, rain is necessary. After irrigation or rainfall, the pores are mainly filled with water. If no additional water is supplied to the soil, it gradually dries out. For farming, soil quality plays a significant role in crop productivity since soil nutrients and soil physical properties can directly impact yields. Some living material is also found in the soil like beetles, worms, larvae, etc. They help to aerate the soil and thus create favorable growing conditions for the crop.

Thimme et al. (2013) defined crop as "Aggregation of individual plant species grown in a unit area for economic purpose" in his crop growth model. Fleming et al. (2012) studied on deterministic and stochastic optimal control. The mathematical theory of optimal process is formulated by Pontriag in et al. (1986). Shah et al. (2017) published papers on optimal control on depletion of green belt due to industries and optimum control for spread of pollutants through forest resources. Greaves and Wang (2016) used statistical way to assess the AquaCrop model for simulating Maize growth in a tropical environment. Using multi-objective fuzzy–robust programming (MOFRP), Qian et al. (2017) has developed model for optimal use of agricultural water and land resources under socio-economic and ecological objectives. Kumar and Chaturvedi (2009) illustrated crop modeling as a tool for agricultural research. For simulating efficiently water-limited crop production, Foster et al. (2017) gave open-source version of the FAO aqua crop model. Brouwer et al. (1985) gave theory on importance of soil and water. Delécolle et al. (1992) used remote sensing as a tool for estimating crop production on a regional scale. Recently, Rendezvous (2018) explored survey for importance of crop production in India. Diekmann et al. (1990) computed the basic reproduction ratio R_0 for infectious disease in heterogeneous population. For stability was discussed using Lyapunov function. Hale (1980) and Lasalle (1976) discussed the general theory of Lyapunov function.

In Section 2, we illustrate a mathematical model. It is a schematic representation of a set of nonlinear differential equations, which represents behavior of a system. The transmission diagram will indicate the process. For growth of crop, we consider the state variables in the proposed model. The threshold value is calculated to study the system. In Section 3, the stability analysis is considered. By deriving its numerical simulation in Section 4, observations are listed in Section 5. Through the model, one can predict the changes in crop status with respect to time.

8.2 Mathematical Modeling

In this paper, the model consists of water volume (W) dissolve with enough density of soil (S) which converts into fertile soil (F_S) and then results into crop production (C). Also, here we have considered that crop watered directly with some rate.

Table 8.1 Notations and parametric values

Notation		Parametric Values
B	The growth rate of water volume	0.40
β	The rate at which soil absorbs water	0.30
δ	The increase rate of fertile soil	0.80
ε	The rate at which crop yield increases due to fertile soil	0.80
η	The rate at which crop yield increases due to water	0.60
μ	Wastage rate of variable from respective compartment	0.55

The notations and parametric values used in dynamical system of the model are given in the following Table 8.1.

With the help of above model parameters, the mathematical model is formulated with the help of the transmission diagram shown in Figure 8.1.

The system of non-linear differential equations for the model is as given below:

$$\frac{dW}{dt} = B - \beta WS - \eta WC - \mu W$$
$$\frac{dS}{dt} = \beta WS - \delta S - \mu S \tag{8.1}$$
$$\frac{dF_s}{dt} = \delta S - \varepsilon F_s - \mu F_s$$
$$\frac{dC}{dt} = \eta WC + \varepsilon F_s - \mu C[-2ex]$$

where $W + S + F_s + C = N$. Also, $W > 0; S, F_s, C \geq 0$

Adding the above differential equations of system (8.1), we have

$$\frac{d}{dt}(W + S + F_s + C) = B - \mu W - \mu S - \mu F_s - \mu C$$
$$= B - \mu(W + S + F_s + C) \geq 0$$

which implies that $\lim_{t \to \infty} \sup (W + S + F_s + C) \leq \frac{B}{\mu}$.

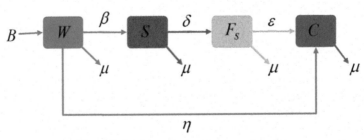

Figure 8.1 Transmission diagram of model.

Therefore, the feasible region of the model is

$$\Lambda = \left\{ (W, S, F_s, C) \in R^4 : W + S + F_s + C \leq \frac{B}{\mu} \right\}.$$

Now, the equilibrium point of the transmission of model is $E_0 = (\frac{B}{\mu}, 0, 0, 0)$.

Now, we calculate the threshold called basic reproduction number R_0 using next generation matrix method.

Let

$$X' = (W, S, F_s, C)' \text{ and } X' = \frac{dX}{dt} = F(X) - V(X).$$

where $F(X) =$ is for the rate of appearance of new individual in component and $V(X) =$ is for the rate of transfer of culture which is given by

$$F(X) = \begin{bmatrix} \beta W S \\ \eta W C \\ 0 \\ 0 \end{bmatrix} \text{ and } V(X) = \begin{bmatrix} \delta S + \mu S \\ -\varepsilon F_s + \mu C \\ -B + \beta W S + \eta W C + \mu W \\ -\delta S + \varepsilon F_s + \mu F_s \end{bmatrix}$$

Now, $DF(X_0) = \begin{bmatrix} f & 0 \\ 0 & 0 \end{bmatrix}$ and $DV(X_0) = \begin{bmatrix} v & 0 \\ J_1 & J_2 \end{bmatrix}$

where f and v are 4×4 matrices defined as

$$f = \left[\frac{\partial F_i(X_0)}{\partial X_j} \right] \text{ and } v = \left[\frac{\partial V_i(X_0)}{\partial X_j} \right]$$

Finding f and v, we get

$$f = \begin{bmatrix} \beta W & 0 & \beta S & 0 \\ 0 & \eta W & \eta C & 0 \\ 0 & 0 & 0 & 0 \\ 0 & 0 & 0 & 0 \end{bmatrix} \text{ and } v = \begin{bmatrix} \delta + \mu & 0 & 0 & 0 \\ 0 & \mu & 0 & -\varepsilon \\ \beta W & \eta W & \beta S + \eta C + \mu & 0 \\ -\delta & 0 & 0 & \varepsilon + \mu \end{bmatrix}$$

Here, v is non-singular matrix.

Therefore, the expression of basic reproduction number R_0 is as below:

$$R_0 = \text{Spectral radius of matrix } fv^{-1} = \frac{\beta B \mu(\varepsilon + \mu) + \eta B \varepsilon \delta}{\mu^2(\delta + \mu)(\varepsilon + \mu)} \tag{8.2}$$

8.3 Stability Analysis

In this section, local and global stability of the equilibrium points are discussed. We have event free equilibrium and endemic equilibrium points.

8.3.1 Local Stability

First, we calculate the local stability behavior of equilibrium $E_0 = \left(\frac{B}{\mu}, 0, 0, 0\right)$ by using Jacobian matrix J_0. The Jacobian matrix of the model is as given below:

$$
J_0 =
\begin{bmatrix}
-\mu & \frac{-\beta B}{\mu} & 0 & \frac{-\eta B}{\mu} \\
0 & \frac{\beta B}{\mu} - \delta - \mu & 0 & 0 \\
0 & \delta & -\varepsilon - \mu & 0 \\
0 & 0 & \varepsilon & \frac{\eta B}{\mu} - \mu
\end{bmatrix}
$$

Above Jacobian J_0 has eigenvalues,

$$\lambda_1 = -\mu$$
$$\lambda_2 = -\varepsilon - \mu$$
$$\lambda_3 = \frac{-\mu^2 + B\eta}{\mu}$$
$$\lambda_4 = -(\mu^2 + \delta\mu - B\beta)$$

If all eigenvalues of Jacobian matrix are negative, then equilibrium point of system is stable. Here, third eigenvalue is positive as all values of parameters are positive. So, we derive one condition as below.

$$-\mu^2 + B\eta < 0$$

for the equilibrium point to be stable.

Next, we determine the local stability behavior of equilibrium $E^* = (W^*, S^*, F_S^*, C^*)$ by using the Jacobian matrix J with

$$W^* = \frac{\delta + \mu}{\beta},$$

$$S^* = \frac{(B\beta - \mu(\delta + \mu))(\eta(\delta + \mu) - \beta\mu)(\varepsilon + \mu)}{\mu\beta((\delta + \mu)(\eta\varepsilon - \beta\varepsilon - \beta\mu + \eta\delta + \eta\mu)},$$

$$F_S^* = \frac{\delta(B\beta - \mu(\delta + \mu))(\eta(\delta + \mu) - \beta\mu)}{\mu\beta((\delta + \mu)(\eta\varepsilon - \beta\varepsilon - \beta\mu + \eta\delta + \eta\mu)},$$

$$C^* = \frac{\delta\varepsilon(-B\beta + \mu(\delta + \mu))}{\mu((\delta + \mu)(\eta\varepsilon - \beta\varepsilon - \beta\mu + \eta\delta + \eta\mu))}.$$

The Jacobian matrix of the model is given by:

$$J^* = \begin{bmatrix} -\beta S^* - \eta C^* - \mu & -\beta W^* & 0 & -\eta W^* \\ \beta S^* & \beta W^* - \delta - \mu & 0 & 0 \\ 0 & \delta & -\varepsilon - \mu & 0 \\ \eta C^* & 0 & \varepsilon & \eta W^* - \mu \end{bmatrix}$$

The corresponding characteristic equation of Jacobian matrix is,

$$\lambda^4 + a_1\lambda^3 + a_2\lambda^2 + a_3\lambda + a_4 = 0$$

where

$$a_1 = (-\beta W^* + \delta + \mu) + (-\eta W^* + \mu) + 2\mu + \beta S^* + \eta C^* + \varepsilon$$

$$\begin{aligned} a_2 = {} & (-\beta W^* + \delta + \mu)(\eta C^* + 3\mu + \varepsilon) + (-\eta W^* + \mu)(\beta S^* + \varepsilon + 3\mu) \\ & + \eta W^*(W^*\beta - \delta) + \delta\beta S^* + 2\mu\beta S^* + 2\mu\eta C^* + \varepsilon\beta S^* + \varepsilon\eta C^* + \varepsilon\mu \end{aligned}$$

$$\begin{aligned} a_3 = {} & (-\eta W^* + \mu)(\varepsilon\delta + 2\varepsilon\mu + 2\delta\mu + \varepsilon\beta S^* + 2\mu\beta S^* + \beta S^*\delta + 3\mu^2) \\ & + (-\beta W^* + \delta + \mu)(\varepsilon\eta C^* + \mu\varepsilon + 2\mu\eta C^* + \mu^2) - \beta\mu^2 \\ & (-2W^* + S^*) + \mu\varepsilon\beta(-W^* + S^*) + \eta W^{*2}\varepsilon\beta + 2\eta W^{*2}\beta\mu \\ & + \mu\varepsilon\eta C^* + \mu\beta S^*\delta + \varepsilon\beta S^*\delta + \mu^2\eta C^* \end{aligned}$$

$$\begin{aligned} a_4 = {} & (-\eta W^* + \mu)(\varepsilon\mu\beta S^* + \mu\delta\beta S^* + \varepsilon\mu^2 + \mu^2\delta + \varepsilon\delta\mu + \mu^2\beta S^* + \mu^3 \\ & + (-\beta W^* + \delta + \mu)(\varepsilon\mu\eta C^* + \mu^2\eta C^*) + \varepsilon\mu\beta(-\mu W^* + \delta S^*) \\ & - \mu^3\beta W^* + \eta W^{*2}\mu^2\beta + \eta W^{*2}\varepsilon\beta\mu \end{aligned}$$

The equilibrium point is locally asymptotically stable, if following conditions hold:

(a) $-\eta W^* + \mu > 0$
(b) $-\beta W^* + \delta + \mu > 0$
(c) $S^* > W^*$

A last condition tells that, density of soil should be more than that of water.

8.3.2 Global Stability

Here we discuss the global stability behavior of the equilibrium E_0 and E^* by Lyapunov function. First, we study the behavior of the global stability of E_0.

Consider the Lyapunov function

$$L(t) = S(t) + F_s(t) + C(t)$$
$$L'(t) = S'(t) + F_s'(t) + C'(t)$$
$$= \beta W S + \eta W C - \mu S - \mu f_l - \mu C$$
$$= \left(\frac{\beta B}{\mu} - \mu\right) S + \left(\frac{\eta B}{\mu} - \mu\right) C - \mu F_s$$

$L'(t) < 0$ if $\frac{B\beta}{\mu} - \mu < 0$ and $\frac{B\eta}{\mu} - \mu < 0$

If $B\beta < \mu^2$ and $B\eta < \mu^2$ then E_0 is globally stable.
Next, we study the behavior of the global stability of E^*.
Consider the Lyapunov function,

$$L(t) = \frac{1}{2}[(W - W^*) + (S - S^*) + (F_s - F_s^*) + (C - C^*)]^2$$
$$L'(t) = [(W - W^*) + (S - S^*) + (F_s - F_s^*) + (C - C^*)]$$
$$(W' + S' + F_s' + C')$$
$$= [(W - W^*) + (S - S^*) + (F_s - F_s^*) + (C - C^*)]$$
$$(B - \mu(W + S + F_s + C))$$
$$= [(W - W^*) + (S - S^*) + (F_s - F_s^*) + (C - C^*)]$$
$$(-\mu((W - W^*) + (S - S^*) + (F_s - F_s^*) + (C - C^*)))$$
$$= -\mu[(W - W^*) + (S - S^*) + (F_s - F_s^*) + (C - C^*)]^2 \leq 0$$

Here, we denote $B = \mu(W^* + S^* + F_s^* + C^*)$
   Therefore, E^* is globally stable.

Theorem: E^* is globally stable.

8.4 Numerical Simulation

In this section using the values given in Table 8.1, numerical simulation will
be done.
   Figure 8.2 shows the effect of each compartment. It can be observed that
soil needs water in 0.05 years so that after 0.45 years it becomes fertile soil
which results 15.55% growth in crop production. Also see that after one year
amount of water should be increased, then and only then growth of crop will
be increased. Otherwise less amount of water may be harmful to density of
fertile soil or crop may fail.

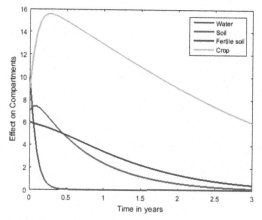

Figure 8.2 Effect on all compartments.

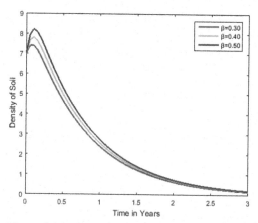

Figure 8.3 Density of soil for different value of β.

From Figure 8.3, one can analyze the effect of the rate at which soil absorbs water. As the value of β is increased from 0.30 to 0.50 the density of soil also increased by 10.5%. It is advised to maintain water level, as both are important actor for crop yield.

The effect of increase rate of fertile soil can be studied from Figure 8.4. One can observe that density of fertile soil decreases gradually with the rate of δ. The density of soil and the water volume should be in control to maintain the density of fertile soil. From Figure 8.5, we can conclude the rate at which crop increases due to fertile soil also affected positively on crop. Increasing the value of ε also increases the growth of crop, is increased by 3.09%.

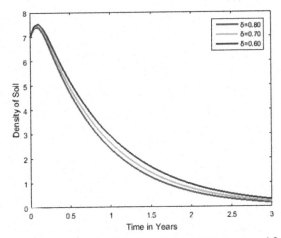

Figure 8.4 Density of fertile soil for different value of δ.

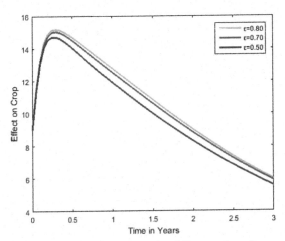

Figure 8.5 Effect on crop for different value of ε.

Figure 8.6 gives the effect of rate at which crop increases due to water. As we increase the value of η, the growth rate of crop will also increase. The value of η increased from 0.60 to 0.80 gives increment in growth of crop by 4.7%. It shows the positive effect on crop, but after half of year as we show previous conclusion that amount of water which provided directly to crop should be increased.

Figure 8.7 represents the pie chart of simulating model. Shortage of water is worldwide issue which can be noticed here that it is less consumed as 3%

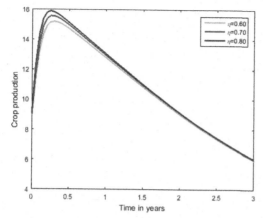

Figure 8.6 Effect of η on crop.

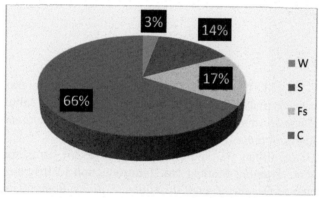

Figure 8.7 Percentage of crop yield by water and soil.

only. 14% soil absorbs water which results into 17% of fertile soil. This fertile soil will help to increase the growth of crop production by 66%.

8.5 Conclusion

In this chapter, an application of *SEIR* model is studied for the crop through water and soil texture. Study reflects the behavior of the crop yield that occurs when crop comes in the contact of other model parameters. Mathematical model is developed using non-linear differential equations which have written to describe a process for crop production under certain combinations of circumstances. The system gives equilibrium points, from which stability is carried out. The stability of model gives conditions from which it is observe that density of soil should be more than the amount of water. Numerical simulation of the model states that through density of soil should be more; water should be increased to maintain density of fertile soil. It also advised that large amount of water is harmful for fertile soil consequently for the crop yield. Therefore, better water management and density of soil texture will give us huge amount of crop produce.

Using the value of parameters given in Table 8.1, the basic reproduction number is calculated as 0.4402 which indicates that 44.02% crop is grown when water level is maintained.

Acknowledgement

The authors acknowledge DST-FIST file # MSI-097 for technical support to the department.

References

[1] Abdulaziz, A. M. (2010). A mathematical modeling approach for irrigation water management under water shortage and salinity conditions: The wave_ms model. In *Proc. Fourteenth International Water Technology Conference, IWTC* (Vol. 14, p. 2010).

[2] Brouwer, C., Goffeau, A., and Heibloem, M. (1985). Irrigation Water Management: Training Manual No. 1-Introduction to Irrigation. *Food and Agriculture Organization of the United Nations, Rome, Italy.*

[3] Delécolle, R., Maas, S. J., Guérif, M., and Baret, F. (1992). Remote sensing and crop production models: present trends. *ISPRS Journal of Photogrammetry and Remote Sensing, 47*(2–3), 145–161.

[4] Diekmann, O., Heesterbeek, J. A. P., and Metz, J. A. (1990). On the definition and the computation of the basic reproduction ratio R 0 in models for infectious diseases in heterogeneous populations. *Journal of mathematical biology, 28*(4), 365–382.

[5] Foster, T., Brozović, N., Butler, A. P., Neale, C. M. U., Raes, D., Steduto, P., and Hsiao, T. C. (2017). AquaCrop-OS: An open source version of FAO's crop water productivity model. *Agricultural water management, 181*, 18–22.

[6] Greaves, G. E., and Wang, Y. M. (2016). Assessment of FAO AquaCrop model for simulating maize growth and productivity under deficit irrigation in a tropical environment. *Water*, *8*(12), 557.

[7] Hale, J. K. Ordinary differential equations. 1980. *RE Krieger Publ.*

[8] http://ec.europa.eu/environment/soil/pdf/Soil%20and%20Water.pdf

[9] http://www.rroij.com/open-access/crop-growth-modeling-a-review-1-11.pdf

[10] https://www.mbarendezvous.com/essay/importance-of-crops-in-india/

[11] https://en.wikipedia.org/wiki/Compartmental_models_in_epidemiology

[12] https://en.wikipedia.org/wiki/Basic_reproduction_number

[13] https://en.wikipedia.org/wiki/Epidemiology

[14] Kumar, R., and Chaeturvedi, S. (2009). Crop modeling: A tool for agricultural research.

[15] LaSalle, J. P. (1976). *The stability of dynamical systems* (Vol. 25). Siam.

[16] Ojo, M. M., and Akinpelu, F. O. Lyapunov Functions and Global Properties of SEIR Epidemic Model.

[17] Pontriagin, L. S., Boltyanskii, V. G. Gamkrelidze, R. V. and Mishchenko, E. F. (1986). The Mathematical theory of Optimal Process. Gordon and Breach Science Publishers, New York, 4–5.

[18] Shah, N. H., Satia, M. H., and Yeolekar, M. (2017). Optimal Control on depletion of Green Belt due to Industries. Advances in Dynamical Systems and Applications, *12*(3), 217–232.

[19] Shah, N. H., Satia, M. H., and Yeolekar, B. M. (2017). Optimum Control for Spread of Pollutants through Forest Resources. Applied Mathematics, *8*(5), 607.

[20] Tan, Q., Zhang, S., and Li, R. (2017). Optimal use of agricultural water and land resources through reconfiguring crop planting structure under socioeconomic and ecological objectives. *Water*, *9*(7), 488.

[21] Zingaro, D., Portoghese, I., and Giannoccaro, G. (2017). Modeling crop pattern changes and water resources exploitation: A case study. *Water*, *9*(9), 685.

9

Advances in Radial Basis Functions

Geeta Arora[*] **and Gurpreet Singh Bhatia**

Lovely Professional University, Phagwara, Punjab, India
E-mail: geetadma@gmail.com; gurpreetsidakbhatia@gmail.com
*Corresponding Author

This chapter is an attempt to summarize and understand the developments of radial basis function in terms of application to other legacy numerical methods. Most of the development of this method lies in the choice of shape parameter and the type of basis function involved in the implementation of that method. In this chapter, the methodology for Kansa method and symmetric collocation method using radial basis function is explained in brief along with the detail discussion of radial basis function pseudospectral method (RBF-PS) approach. Implementation of the method is presented to solve Chaffee-Infante (CI) equation with numerical examples. The obtained results are depicted in form of tables and figures.

9.1 Introduction

With the advancement of technology, it has become possible to depict a relationship between two or more than two entities involved in the various phenomenon of science by the application of mathematical modelling. The process of mathematical modelling can be used to express any phenomenon in term of differential equations which can be either an ordinary or a partial differential equation. Much of the work reported so far during the last decades has been done to find the solution for these obtained equations. However, until now no unique analytical technique could be established. However, numerical methods can serve as an effective tool in solving complex differential equations due to ease of implementation in comparison to analytical approaches.

Many of the techniques or schemes that have proved their efficiency in handling the ordinary differential equation (ODE) include but not limited to Picard's method, Taylor's series method, and R.K. method. Similarly, various well-known methods such as finite difference method, finite element method, differential quadrature method, variation iteration method, adomian decomposition method, etc. are developed in recent years to solve partial differential equations (PDE). Although, these developed methods are able to find the solution of the various equations but are still affected by factors such as slow rate of convergence, sensitivity to the initial approximations or high computational cost.

Numerical methods based on radial basis function approach are well-known, as they are easy to implement with uniform or non-uniform mesh in a programmable approach and provides accurate solutions. Thus, in the present scenario, numerical methods using radial basis functions remain a favourite topic of researchers to find the solution of PDEs. Though a lot of work is being done to find solutions to some of the recognized PDEs such as Fisher's equation, Burgers' equation, etc. but the research still remains in its initial stage providing researchers both opportunity and challenge to develop an algorithm to estimate parameters related to radial basis functions.

9.2 Requirement of Mesh-free Methods

Many areas of science and engineering require data in a high dimensional form. To deal with such type of requirements, classical numerical methods with uniform discretization process are not suitable; this leads to the development of mesh free techniques. These mesh-free methods were developed to deal with the flexible geometry of problems in the area of interest. These methods are also very efficient in terms of computational cost due to the association of independent points as grid or node points for the computational work.

Initially, the mesh-free approximation methods have been applied by the researchers in the area related to metrology, geophysics and geodetics. Later on, these methods of approximation have been applied in various other emerging areas of research. Some of the application areas where mesh free methods play an important role include scattered data modelling [1], the solution of partial differential equations in higher dimensions [2], in mathematical finance [3], in computer graphics [4], in optimization [5] and also in neural networks [6].

Furthermore, mesh-free methods are known as new generation numerical tools. This name itself implies their capability to handle multiple parameters involved in the application problem in contrast to the classical numerical methods which can handle problems only in two or three dimensions (parameters).

9.3 Radial Basis Function

The Gaussian function is a well-known function that occurs in many of the branches of mathematics given by:

$$\phi(r) = e^{-(\varepsilon r)^2}, \quad r \in \mathbb{R}.$$

Here, ε is a parameter that controls the shape of the generated curve called as shape parameter and r represents the distance of node points either from the origin or from the center points.

Definition:

Radial basis function (RBF) can be defined as a function which is univariate, real-valued and continuous. Value of an RBF depends upon the distance of node points from the origin [7]. Mathematically, a function $\chi : \mathbb{R}^n \to \mathbb{R}$ is a radial basis function if it satisfies the following:

$$\chi(x) = \phi(r), \quad r = \|x\| \in \mathbb{R}$$

Here, $\|\bullet\|$ represents the Euclidean norm, which is a measure of length for a Vector^x. In other words, it is the distance of a node from the origin or from some other node points, called as a center. Thus r can also be written as: $r = \|x - x_k\|$ for some node point x_k.

9.4 Properties of Radial Basis Functions

1. Invariance under transformations such as translation, rotation, and reflection.
2. Independence of the complexity involves in approximating multivariate function.
3. Flexibility with the geometry of the problem.
4. Has a high rate of convergence.
5. Has compact support which results in sparse interpolation matrix.

9.5 Developments in Radial Basis Functions

Definition of radial basis functions was given by Hardy [13] in the context of its application to topology in form of a multi-quadric function which is a globally supported interpolant. This method has been verified by many researchers for parameters such as accuracy, CPU time and memory requirement. In 1982, Franke [14] in his work presented the superiority of these functions to solve interpolation problem of scattered data. This function was further used by Kansa [15–17] in early 90s to solve a partial differential equation. Till then this method has gone through various stages of development but become famous as Kansa method. Despite of its various advantages over other techniques, this method is affected from the ill-conditioning of the generated matrix for a large number of nodes. This problem of ill-conditioning was further taken care by Fasshauer [18] by proposing the concept of symmetric RBF collocation approach. Furthermore advancement has been done by various researchers to deal with the problem related to the ill-conditioned matrix using preconditioning of matrix, domain decomposition method, a local approach based on RBF etc. All this lead to the generation of multiple forms of RBF [7], thus development of Gaussian function (GS), linear radial function (LR), multi-quadric (MQ), and inverse multi-quadric (IMQ). Following is a table with discussed formulae of some important radial basis functions:

Name of the RBF	Formula, $\phi(r) =$
Gaussian Function (GS)	$e^{-(\varepsilon r)^2}$
Linear radial function (LR)	r
Multi-quadric (MQ)	$\sqrt{1 + (\varepsilon r)^2}$
Inverse quadric (IQ)	$\frac{1}{1+(\varepsilon r)^2}$
Inverse Multi-quadric (IMQ)	$\frac{1}{\sqrt{1+(\varepsilon r)^2}}$

9.6 Shape Parameters

In the above-mentioned approaches, there exists a parameter ε, called as shape parameter that plays an important role in RBF. A small change in the value of shape parameter can bring a major change in terms of accuracy. From last several years, researchers have proposed different approaches to find this shape parameter. Some of them are as follows:

(a) **Hardy shape parameter:** According to Hardy [13], the shape parameter can be calculated as the reciprocal of another parameter c which is 0.815 times the average distance (d) between nodes. If N is the total number of center nodes, then the average distance can be calculated as: $d = \frac{1}{N} \sum_{k=1}^{N} d_k$. Thus shape parameter is given by

$$\varepsilon = \frac{1}{c}, \quad c = 0.185d$$

(b) **Franke shape parameter:** Franke [14] also considered shape parameter value as reciprocal of parameter c where parameter c depends on the diameter D of the smallest circle enclosing all data points. Thus, shape parameter is defined as:

$$\varepsilon = \frac{1}{c}, \quad \text{with} \quad c = \frac{0.125D}{\sqrt{N}}$$

(c) **Exponentially varying shape parameter:** This approach has been proposed by Kansa [15–16]. According to Kansa, the value of shape parameter depends upon the maximum and minimum values of shape parameter given by:

$$\varepsilon_{\min} = \frac{1}{\sqrt{N}} \quad \text{and} \quad \varepsilon_{\max} = \frac{3}{\sqrt{N}}.$$

(d) **Fasshauer's variable shape parameter strategy:** Fasshauer [18–19] proposed that the value of shape parameter must not be unique for the whole domain but can vary corresponding to each center. This result in different values of shape parameter for each column of RBF coefficient matrix. According to him, shape parameter can be written in the form given by:

$$\varepsilon = \frac{\sqrt{N}}{2}$$

(e) **Rippa's optimal approach:** Rippa [20] proposed an algorithm based on the optimization approach, also known as leave-one-out cross validation (LOOCV). According to him, the value of shape parameter can be calculated by minimizing the value of cost function taken as a function of RMS error calculated between numerical and exact solution. The value of shape parameter is given by the relation

$$\varepsilon_i(c) = \frac{u_i}{A_{ii}^{-1}}$$

Here, A_{ii}^{-1} is the diagonal element of the inverse of the coefficient matrix and u_i is a component of vector u.

(f) **Random variable shape parameter:** As proposed by Sarra and Sturgill [21], the value of shape parameter can be calculated based upon the maximum and minimum values of shape parameter given by:

$$\varepsilon_j = \varepsilon_{\min} + (\varepsilon_{\max} - \varepsilon_{\min})rand(1, N)$$

Here rand( ), is a MATLAB function to generate uniformly distributed random numbers on unit interval.

(g) **Gherlone's approach:** Gherlone [22] developed an algorithm to calculate optimal shape parameter based on convergence analysis. The optimization is based on the rate of convergence, for the solution obtained with N_{k+1} and N_k nodes. The algorithm is to minimize the rate of convergence for given choice of shape parameter, given by:

$$r = \left| \frac{u(N_{j+1}, c_k) - u(N_j, c_k)}{N_{j+1} - N_j} \right|$$

where, $u(N_i)$ denoted the solution estimate with N_i nodes and a value c_k of shape parameter.

(h) **Trigonometric variable shape parameter:** This approach has been proposed Xiang et al. [23]. They proposed, that the value of shape parameter depend upon its maximum and minimum values and gave a formula similar to random variable shape parameter [9] corresponding to each column using values of sin given by:

$$\varepsilon_j = \varepsilon_{\min} + (\varepsilon_{\max} - \varepsilon_{\min})sin(j), \quad j = 1 \text{ to } N$$

(i) **Global and Local Optimization using Direct Search (GLODS):** Roque [24] in 2017 proposed the application of an optimization technique known as Global and Local Optimization using Direct Search (GLODS) in optimizing shape parameter of the multiquadric function. He discussed the results obtained using this algorithm in application of Radial Basis Function-Finite Difference (RBF-FD).

9.7 RBF Interpolation

Interpolation is a well-known method used in various fields for predicting the value of a function at some points of a data set when the value of the function is known for some given points in the same data set. There are several numerical methods for the interpolation of data such as Hermite interpolation, Spline interpolation, etc. But when the data is not uniformly distributed, RBF interpolation is the most preferred choice.

Let us consider a problem of finding a function $p(x)$ for the given set of data points (x_k, y_k) with $k = 1$ to N, $x_k \in \mathbb{R}^n$, $y_k \in \mathbb{R}$. We have considered the points $x_k \in \mathbb{R}^n$ to get freedom in choice of the dimension of domain points. If $n = 1$, data represents a series of values measured over a period of time. If $n = 2$, data corresponds to two values of a coordinate plane and for $n = 3$, data points imply the view of three-dimension coordinates. As an example, one can consider the measurement of velocity gain by a rocket after launch in an interval of seconds representing data in one-dimension. For a two-dimension, one can consider the hike in temperature measured during summer at different parts of our country. The position of a space shuttle in space recorded over the different time interval is an example of three dimensional data.

For interpolation with uniform data values, the domain can be discretized using a common space step, but to deal with the randomly scattered data values, Halton points are considered which are the uniformly distributed random points in interval $[0, 1]$.

For the problem in one-dimension, if the data points are uniformly distributed, then splines are the best alternative as they have been developed to deal with the complexity involved in the interpolation process of formulating a polynomial with an increase in the number of data points. They are the piecewise polynomial that is nonzero for a given interval that otherwise has the value zero at other points of the domain. After spline, the best alternative to deal with the scattered data points is radial basis function. Let us discuss the process with an example:

To obtain an approximate interpolation function, consider a function of form

$$p(x) = \sum_{k=1}^{n} c_k \phi \left(\|x - x_k\| \right), \quad x \in \mathbb{R}^n$$

Let the value of the function, known at some given points is given by $f(x)$. On expanding the function, leads to a system of linear equation with coefficient matrix A, with elements $\phi \left(\|x_j - x_k\| \right)$, with c_k as unknown and the column vector y_k, for $k = 1$ to N, given by:

$$
\begin{bmatrix}
\phi(\|_1 - _1\|) & \phi(\|_1 - _2\|) & \cdots & \phi(\|_1 - _N\|) \\
\phi(\|_2 - _1\|) & \phi(\|_2 - _2\|) & \cdots & \phi(\|_2 - _N\|) \\
\vdots & \vdots & \cdots & \cdots \\
\phi(\|_{N-1} - _1\|) & \phi(\|_{N-1} - _2\|) & & \phi(\|_{N-1} - _N\|) \\
\phi(\|_N - _1\|) & \phi(\|_N - _2\|) & \cdots & \phi(\|_N - _N\|)
\end{bmatrix}
\begin{bmatrix}
c_1 \\ c_2 \\ \vdots \\ c_{N-1} \\ c_N
\end{bmatrix}
=
\begin{bmatrix}
f(_1) \\ f(_2) \\ \vdots \\ f(_{N-1}) \\ f(_N)
\end{bmatrix}
$$

For solution of the system to exist, the coefficient matrix must be singular [7]. Hence the unknown can be calculated as solution to the system given by $A^{-1}f$. That can be used to find the value at some other node points.

Following is the algorithm to interpolate data at given points using the Gaussian RBF:

1. Choose an RBF function for the interpolation process.
   Let us choose Gaussian radial basis function given by $e^{-(\varepsilon r)^2}$
2. Define the test function, a function whose value is known at some given points, named as centers.
   Let us take $0.75e^{-((9x-2)^2+(9y-2)^2)/4}$ as a test function, which is a part of famous Franke's function usually taken for scattered data interpolation.
3. For calculating the coefficient matrix, we need the data points as grid with norm of radial function as element.
4. Calculate the right hand side column value by using the value of function at the known points.
5. Find the unknown from the matrix inversion and multiply the coefficients with points at which interpolation value is required.
6. To verify the applicability of the procedure calculate the value of function directly by substitution of node points and then compare it with the product of unknowns to get the error involved.

9.8 Solution of Differential Equation

There are lot of approaches that are developed to find solution of differential equations using radial basis functions. Some of the approaches are briefly discussed below with description of RBF-PS method.

9.8.1 Kansa Method

To implement the Kansa method for the solution of a differential equation, whole domain is firstly discretized into a number of points. The points on the boundary are called as boundary points. The solution of differential equation can be written as a linear combination of RBFs at the predefined points called as nodes, thus given by:

$$u(x) = \sum_{k=1}^{N} \alpha_k \varphi \left(\|x - x_k\| \right)$$

Here N is the number of node points in the domain and $\alpha_k's$ are the unknown coefficient to be obtained.

This method has been used to solve various well-known differential equations such as heat equation, electrostatic equation, shallow water equation and fractional diffusion equation. As discussed above, this method suffers from the problem of an ill-conditioned matrix that leads to more computational cost as the number of nodes is increased.

9.8.2 Symmetric Collocation Method

To overcome the problems of Kansa approach, a symmetric collocation method based on hermite interpolation concept was proposed by Fasshauer. He proposed the concepts of centers. Centers are those interior points from which the distance of all other nodes is calculated to approximate the function. In this approach, the solution is approximated as a sum of the linear combination of two functions.

This leads to the generation of a symmetric matrix. This approach has been proved better by researchers as compared to Kansa method in terms of computational cost. Many of the problems related to the differential equation have been solved using different RBF approaches that includes 2D elastostatic problem and time-dependent PDEs [8, 10, 11].

9.8.3 RBF-pseudospectral Approach

In this section we are going to discuss the radial basis function based pseudospectral (RBF-PS) approach to find numerical solution of Chaffee-Infante (CI) equation [25], a form of reaction-diffusion equation given by:

$$u_t = u_{xx} + \gamma u(1 - u^2) \tag{9.1}$$

Here, γ is a non-negative arbitrary constant which maintain the corresponding balance of the diffusion term u_{xx} and the non-linear term in the equation. When $\gamma = 1$, this equation reduces to the well-known Newell-Whitehead equation, which exists to study the influence of the diffusion term with the involved nonlinear reaction term. It can also be considered as a general case of Allen-Cahn equation which is given by: $u_t = \eta u_{xx} + u(1 - u^2)$ that can be deduced by substituting value of η parameter as one. This equation exists in the modelling of phenomenon related to modelling of quantum and plasma physics, mathematical biology and image processing.

The CI equation is to be solved with required initial and boundary conditions given by:

$$u(x,0) = \varphi(x), \ x \in [a, b] \tag{9.2}$$

$$u(a, t) = u(b, t) = \beta \tag{9.3}$$

The boundary condition given by β is a small parameter which plays an important role in design and analysis of the solution. The CI equation models a dissipative dynamical system with fixed number of node points. As discussed by Davis [26], CI equation is very sensitive to change in the boundary condition. Author in his work discussed the sensitivity of the solution near zero, with effect on convergence of the solution.

RBF-PS method was proposed by Fasshauer [27] in his work reported to optimize the shape parameter in RBF approximation. The advantage of using RBF-PS as compared to other approaches hybrid with RBF includes that the approximate solution depends upon only some selected grid points. This approach of RBF has recently been used by researches to solve various differential equations. This method provides solution to the equation with optimum accuracy with combination of different forms of radial basis functions. To implement the method, discretize the domain into numbers of domain points also known as nodes. The approach for the discretization could be uniform or non-uniform. For the simplicity we are using the uniform distinct scattered nodes x_k with $k = 1$ to N.

The approximation function can be defined as:

$$u(x,t) = \sum_{k=1}^{N} \alpha_k(t)\varphi\left(\|x - x_k\|\right) \tag{9.4}$$

Here φ are radial basis functions and $\alpha_k(t)$ are the unknown interpolation coefficients. The type of RBF can be chosen as per problem. Here we have used the cubic Matérn radial basis function which is a positive definite given by:

$$\varphi(r) = 15 + 15\varepsilon r + 6(\varepsilon r)^2 + (\varepsilon r)^3 e^{-\varepsilon r}.$$

Equation (9.4) can be written in form of a matrix system given by:

$$u = A\alpha \tag{9.5}$$

where, A is a matrix with elements given by $\varphi\left(\|x_i - x_k\|\right)$ and α's are the coefficients to be determined by using Equation (9.4) as

$$\alpha = A^{-1}u \tag{9.6}$$

If the unknown coefficients are calculated, then to calculate the derivatives one can use Equations (9.5) and (9.6) to obtain the values of first derivative with respect to space. The same process can be followed to calculate the higher order derivatives. For instance, the first order derivative can be calculated as follows:

On differentiating Equation (9.5) and substituting value of α from Equation (9.6), we get $u_x = A_x A^{-1}u$ that can be written in simplified form as:

$$u_x = D_x u \text{ with } D_x = A_x A^{-1}$$

To find the numerical solution of Chaffee–Infante equation, substituting the value of derivatives on the right hand side results in an ordinary differential equation that can be solved by any ODE solver numerical technique.

Choice of shape parameter: As discussed in Section 9.6, there are various strategies adopted by the researchers to finalize the process to obtain the required shape parameter. Here, in this work we have used Rippa's approach [20], which uses the on leave-one out cross validation (LOOCV) modified by Fasshauer and Zhang [27]. This approach is based on obtaining the value of shape parameter to minimize the error.

Example: In order to validate the discussed approach let us consider an example of CI equation $u_t = u_{xx} + \gamma u(1 - u^2)$ for $\gamma = 1$, with given initial condition as:

$$u(x,0) = -0.5 + 0.5\tanh(0.3536x), \ x \in [0, 1]$$

and boundary conditions taken from the exact solution given by:

$$u(x,0) = -0.5 + 0.5\tanh(0.3536x - 0.75t)$$

Solution of the equation is calculated at different time levels. The obtained numerical results are depicted in terms of errors by calculating the L_2 and L_∞

Table 9.1 Calculated errors of the example at different time-levels

Error/t	0.002	0.004	0.006	0.008	0.01
RMS_Error	8.1911e-09	3.1707e-08	1.0473e-07	4.0276e-07	1.6425e-06
L_2 Error	3.7536e-08	1.4530e-07	4.7996e-07	1.8457e-06	7.5267e-06
L_∞ Error	3.5057e-08	8.7536e-08	1.0221e-06	5.6102e-06	2.5273e-05

Table 9.2 Point wise absolute error calculated for the obtained solutions at different time-levels

/t	0.001		0.005		0.009	
	RBF-PS	Zahra [28]	RBF-PS	Zahra [28]	RBF-PS	Zahra [28]
0.1	1.7793E-08	2.448E-04	1.5837E-07	9.687E-04	3.0079E-06	1.492E-03
0.2	1.3153E-08	2.000E-04	9.9767E-08	1.016E-03	1.4144E-06	1.761E-03
0.3	6.6050E-09	1.797E-04	3.3688E-08	9.082E-04	7.1240E-07	1.644E-03
0.4	7.6775E-09	1.594E-04	7.3061E-09	8.052E-04	4.7119E-07	1.465E-03
0.5	8.4637E-09	1.410E-04	8.6622E-08	7.124E-04	9.1947E-07	1.296E-03
0.6	3.1393E-09	1.242E-04	3.5827E-08	6.280E-04	1.0274E-06	1.143E-03
0.7	1.2327E-08	1.091E-04	1.6265E-07	5.520E-04	2.2176E-06	1.003E-03
0.8	2.9332E-09	9.502E-05	1.0553E-07	4.824E-04	2.5065E-06	8.536E-04
0.9	1.1117E-08	8.886E-05	1.5327E-07	3.725E-04	1.4464E-06	6.009E-04

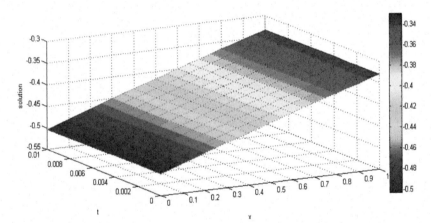

Figure 9.1 Solution of the example obtained at different time-levels.

errors that are enlisted in the form of tabulated results in Table 9.1. The point wise errors are presented in Table 9.2 that also demonstrates the comparison of errors obtained as compared to errors reported in the literature. The physical interpretation of the obtained at the different time-levels is exhibited in the form of Figure 9.1. The value of the shape parameter calculated as per discussion for the calculation of the numerical results is 0.134249.

9.9 Summary

This chapter is an effort to briefly describe the importance of mesh-free techniques using radial basis function approach in solving various complex phenomenon with discussion on the differential equation. It discussed the various issues why numerical methods are more reliable and ease to use as compared to analytical methods.

The radial basis functions are discussed in details along with their properties with emphasis on the development held till now in this field. RBF interpolation is also discussed with discussion on the solution strategies such as Kansa approach, symmetric collocation approach with in-depth study of RBF-pseudospectral approach which is implemented to solve an example of the well-known Chaffee-Infante (CI) equation.

In the implementation of RBF-PS method leave-one out cross validation (LOOCV) approach is used with modification related to the minimization of cost function to obtain the value of shape parameter. The numerical example is solved to calculate solution at various time-levels.

References

[1] Wild S. M., Regis R. G., Shoemaker C. A., ORBIT: Optimization by Radial Basis Function Interpolation in Trust-Regions. SIAM Journal on Scientific Computing. (2008) 30, 6, 3197–3219.

[2] Skala V., RBF Interpolation with CSRBF of Large Data Sets, Procedia Computer Science. (2017) 108, 2433–2437.

[3] Kelly M., Evaluation of Financial Options Using Radial Basis Functions in Mathematica. The Mathematica Journal. (2009) 11, 3 Wolfram Media, Inc.

[4] Kurzendorfer T., Fischer P., Mirshahzadeh N., Pohl T., Brost A., Steidl S. and Maier A., Rapid Interactive and Intuitive Segmentation of 3D Medical Images Using Radial Basis Function Interpolation. J. Imaging (2017) 3, 56.

[5] Wahono B., Praptijanto A., Santoso W. D., Nur A., Suherman and Lu Z., Radial basis function and particle swarm optimization to predict range extender engine performance. Journal of Mechanical Engineering and Sciences. (2017) 11, 4, 3058–3071.

[6] Mosavi M. R., Khishe M., Khani Y. H. and Shabani M., Training Radial Basis Function Neural Network using Stochastic Fractal Search

Algorithm to Classify Sonar Dataset. Iranian Journal of Electrical & Electronic Engineering. (2017) 13, 1, 100–111.

[7] Fssshauer G. E., Meshfree Approximation Methods with MATLAB. World Scientific Publishing Co., Inc. River Edge, NJ, USA ©2007break ISBN: 9812706348 9789812706348

[8] Yensiri S. and Skulkhu R. J., An Investigation of Radial Basis Function-Finite Difference (RBF-FD) Method for Numerical Solution of Elliptic Partial Differential Equations. Mathematics. (2017) 5, 54.

[9] Mavric B., Šarler B., Application of the RBF collocation method to transient coupled thermoelasticity. International Journal of Numerical Methods for Heat and Fluid Flow (2017) 27(5) Emerald Publishing Limited. DOI: 10.1108/HFF-03-2016-0110.

[10] Parand K., Hashemi S., RBF-DQ method for solving non-linear differential equations of Lane-Emden type. Ain Shams Eng J (2016) In press.

[11] Li S., Yao L., Yi S. and Wan W., A Meshless Radial Basis Function Based on Partition of Unity Method for Piezoelectric Structures. Mathematical Problems in Engineering. Hindawi Publishing Corporation (2016) Article ID 7632176, 17 pages.

[12] Saeed R. K. and Sadeeq M. I., Radial Basis Function Pseudospectral Method for Solving Non-Linear Whitham-Broer-Kaup Model. Sohag J. Math. (2017) 4, 1, 13–18.

[13] Hardy R. L., Multiquadratic equations for topography and other irregular surfaces. J Geophys Res. (1971) 176, 1905–1915.

[14] Franke R., Scattered data interpolation tests of some methods. Math Comput. (1982) 38, 181–200.

[15] Kansa E. J., Multiquadrics, A scattered data approximation scheme with applications to computational fluid dynamics I: Surface approximations and partial derivative estimates. Computers and Mathematics with Applications. (1990) 19, 127–145.

[16] Kansa E. J., Multiquadrics, A scattered data approximation scheme with applications to computational fluid-dynamics II: Solutions to parabolic, hyperbolic and elliptic partial differential equations. Computers and Mathematics with Applications. (1990) 19, 147–161.

[17] Kansa E. J. and Carlson R. E., Improved accuracy of multiquadric interpolation using variable shape parameters. Computers and Mathematics with Applications. (1992) 24, 12, 99–120.

[18] Fasshauer G. E., Solving partial differential equations by collocation with radial basis functions, In: Mehaute A., Rabut C., Schumaker L. L.,

editors. Surface Fitting and Multiresolution Methods, Vanderbilt University Press, Nashville. (1997), 131–138.

[19] Fasshauer G. E., Newton iteration with multiquadratics for the solution of nonlinear pdes. Comput Math Appl. (2002) 43, 423–438.

[20] Rippa S., An algorithm for selecting a good value for the parameter c in radial basis function interpolation. Adv. Comput. Math. (1999) 11, 193–210.

[21] Sarra S. A. and Sturgill D., A random variable shape parameter strategy for radial basis function approximation methods. Engineering Analysis with Boundary Elements. (2009) 33, 11, 1239–1245.

[22] Gherlone M., Iurlaro L., Sciuva M., A novel algorithm for shape parameter selection in radial basis functions collocation method. Composite Structures. (2012) 94, 453–461.

[23] Xiang S., Wang K. M., Ai Y. T., Sha Y. D., and Shi H., Trigonometric variable shape parameter and exponent strategy for generalized multiquadric radial basis function approximation. Applied Mathematical Modelling. (2012) 36, 5, 1931–1938.

[24] Roque C. M. C. and Madeira J. F. A., RBF-FD meshless optimization using direct search (GLODS) in the analysis of composite plates, Engineering Analysis with Boundary Elements. (2017) 1–10.

[25] Chafee N. and Infante E. F., A Bifurcation Problem for a Nonlinear Partial Differential Equation of Parabolic Type, Applicable Analysis, (1974) 4, 17–37.

[26] Burns, J. and Davis, L., Sensitivity Analysis and Computational Uncertainty with Applications to Control of Nonlinear Parabolic Partial Differential Equations, Proceedings of the 47th IEEE Conference on Decision and Control. (2008) 3989–3994.

[27] Fasshauer, G. E., and Zhang, J. G., On choosing "optimal" shape parameters for RBF approximation. Numerical Algorithms, (2007) 45(1), 345–368.

[28] Zahra W. K., Trigonometric B-Spline Collocation Method for Solving PHI-Four and Allen–Cahn Equations, Mediterr. J. Math. (2017) 14:12 14: 122. https://doi.org /10.1007/s00009-017-0916-8

10

Modeling for Time Period of Natural Frequency for Non-Homogeneous Square Plate with Variable Thickness and Temperature Effect

Reeta Bhardwaj[1], Naveen Mani[2] and Amit Sharma[1,*]

[1]Department of Mathematics, Amity University Haryana, India
[2]Department of Mathematics, Sandip University, Nashik, Maharashtra, India
E-mail: bhardwajreeta84@gmail.com; naveenmani81@gmail.com;
dba.amitsharma@gmail.com
*Corresponding Author

In this chapter, we applied Rayleigh-Ritz method to study the time period of a non-homogeneous square plate with linear variation in thickness and temperature, on clamped edge condition. For consideration of non-homogeneity, we considered circular variation in density and Poisson's ratio in one dimension. The time period is computed for various values of tapering constant, non-homogeneity constants and thermal gradient. All the results are presented with the help of tables. A comparison of frequency modes with the published results from open literature is also given.

10.1 Introduction

Vibration is the mechanical oscillation of particles (body) from its position of equilibrium or we can say that vibration is the study of the relation between the motions of bodies to the forces acting on them. Generally, in nature, we can say that everything vibrates. We can classify it as weak vibration and strong or disastrous vibration (e.g. tsunamis, earthquake, tornado, etc).

Vibration can be either by human intervention (activities) or natural. Recently, large numbers of research on the study of vibration have been motivated, by the different engineering applications such as the design of automobiles, structures, control systems, wind turbines, and nuclear reactors.

Controlled vibrations are very useful in our daily life such as in case of music, cars, trains, planes, mobiles, electric motors and engines while uncontrolled vibration has a disastrous effect on human life. Therefore, a major important aspect of the study of vibration is to reduce vibration or control the vibration.

Non-homogeneous plates along with non-uniformity have a wide variety of applications in science and engineering such as mechanical engineering, civil engineering, aerospace engineering, and optical engineering because of high strength, light weight, and having higher stiffness. Almost all engineering structures work under the influence of huge temperature, therefore study the vibration with-out consideration of temperature effect is nothing. In order to control the vibration, we have to control the variation in plate parameters.

10.2 Literature Survey

Many researchers have been worked to study the vibrational characteristics of the plate and applied various methods to solve the vibration problem.

The bending (pure and symmetrical) of long plate, circular, anisotropic and plates having lateral loads, forces in the middle has been discussed in [1]. Plate vibration of different shapes (circular, elliptical, rectangular, skewed, anisotropic and plates with variable thickness) on various combination of boundary conditions such as clamped, simply supported and free is studied in [2] based on Classical plate theory. The vibration of the square, rectangle, parallelogram, triangle, and circular plates are discussed and numerical results are presented for homogenous and uniform plates in [3]. Buckling and bending of thin plates and shells are presented in [4]. Excellent research on free transverse vibration of thin plates [5] (homogeneous and isotropic) with a constant thickness, effects of small deflection and no inplane inertial force on vibration have been summarized. A contemporarily relevant literature survey [6] using Mindlin and the modified Mindlin theory, on vibrational analysis of thick plates of different shapes and for laminated plates is presented. A method is presented to show the effects of openings on vibrational frequencies in plates [7] on various boundary conditions. The effect of parabolic thickness on the vibration of the square plate [8] is presented and obtained frequencies and modes shape using finite element analysis. The results are compared with measurements (made with real-time laser holography) and found to be in good agreement except for a few of the lower modes. Non-linear vibration of the orthotropic square plate [9] carrying a concentrated mass on clamped boundary conditions is discussed using Von Kármán's equations. Von Kármán's equations are applied to analyze the large amplitude free vibrations of a square plate [10] of exponentially varying thickness and computed numerical results. Natural vibration of rectangular plate (thin and thick) [11] with arbitrary variable thickness is discussed by using an approximate method, based on the Green function, and numerical solutions for frequency parameter with accuracy and convergence are evaluated. Model characteristics of the rectangular plate [12] with elastic support on edges, by using the Rayleigh-Ritz method are presented. As a result, a large number of numerical results are given to demonstrate the accuracy and convergence of the results. Free vibration of the square plate [13] on a combination of clamped and simply supported edge condition is examined and first two modes of vibration are evaluated on a different variation of non-homogeneity constant, tapering constant and thermal gradient. Two-dimensional linear temperature effects are examined on the natural vibration of the square plate [14] with circular variation in thickness and Poisson's ratio on clamped edge conditions based on Classical plate theory. The effect of variation in density and Poisson's ratio on the clamped square plate [15] with variable thickness (linear) and temperature (bi-parabolic) is presented, and the first two modes are computed. Rayleigh-Ritz method is applied to calculate the vibrational frequency of visco elastic square plate [16] on clamped boundary conditions, with variable Poisson's ratio and thickness along with two-dimensional parabolic temperature variations. Novel separation of variables is used to compute the exact solution for the vibration of rectangular plates (thin and orthotropic) [17] with mixed boundary conditions. The vibrational frequency of homogeneous rectangular plate [18] (isotropic and thin) is studied by using Classical plate theory. A model is constituted to study the natural vibration of the square plate [19] having linear variation in thickness, circular variation in density and two-dimensional variations in temperature using the Rayleigh-Ritz method. Natural vibration of parallelogram plate [20] having circular variation in thickness and Poisson's ratio, with two-dimensional linear temperature variations is studied on clamped edges using the Rayleigh-Ritz method. The effect of two-dimensional thickness, temperature variation and one-dimensional variation in density on the vibrational frequency of skew plate (non-homogeneous) [21] are presented using the Rayleigh-Ritz method on clamped boundary conditions. Rayleigh's method is used to evaluate the fundamental frequency with fixed corners and different mass attachments configurations of thin elastic rectangular, isotropic and orthotropic, plates [22]. Asymmetric vibrations of circular plates [23] with parabolically varying thickness, exponential variation in Young's modulus and density, along the radial direction is discussed using Ritz method on clamped, simply supported and free edges. Natural vibration of moderately rectangular plates (thick and orthotropic) [24] with general boundary restraints, internal line supports and resting on an elastic foundation is presented by using first-order shear deformation theory and new results are obtained subjected to various parameters (elastic boundary restraints, arbitrary internal line supports and resting on elastic foundations). Vibration of circular plates [25] having constant plane load is provided. A Green function is determined for the governing equations and natural frequencies are evaluated. A theoretical analysis of the natural frequency of radial vibration of the circular plate [26] is discussed by using wave propagation approach and classical method to study the radial and piezoelectric effects on the band. Results are also verified by finite element simulation (FEM). An analytical model is studied to analyze the vibration of partially cracked rectangular plates [27] coupled with a fluid medium and evaluated the effect of crack length, fluid length, fluid level, fluid density and immersed depth plate on fundamental frequencies. Free vibration analysis of both thick and thin rectangular plate [28] for various aspect ratios and boundary conditions has been studied using first order shear deformation theory which shows the effect of rotary inertia on the natural frequencies of rectangular plates. An effect of variation in density and two dimensional variations in temperature on the vibrational frequency non-homogeneous square plate [29] with variable thickness are discussed by using the Rayleigh-Ritz technique. Natural frequency of non-homogeneous square plate [30] with thickness and temperature variation, on clamped boundary condition, is studied using the Rayleigh-Ritz technique and first two modes of frequencies are evaluated on a different variation of plate parameters.

In this chapter, authors show the effect of plate parameters especially, circular variation in density, Poisson's ratio and simultaneous variation of both non-homogeneity parameters, on time period of the natural frequency of square plate, on clamped boundary conditions. The effect of thickness and temperature variation on time period of the natural frequency of the plate is also computed. The results are presented with the help of tables.

A comparison of frequency modes of present analysis with existing published results is also given to support the findings of the present study.

10.3 Analysis

The equation of motion for the plate is given by

$$\frac{\partial^2 M_x}{\partial x^2} + 2\frac{\partial^2 M_{xy}}{\partial x \partial y} + \frac{\partial^2 M_y}{\partial y^2} = \rho l \frac{\partial^2 \phi}{\partial t^2}, \tag{10.1}$$

where

$$\left.\begin{array}{l} M_x = -D_1 \left[\frac{\partial^2 \phi}{\partial x^2} + \nu \frac{\partial^2 \phi}{\partial y^2}\right] \\[2mm] M_y = -D_1 \left[\frac{\partial^2 \phi}{\partial y^2} + \nu \frac{\partial^2 \phi}{\partial x^2}\right] \\[2mm] M_{xy} = -D_1 (1 - \nu) \frac{\partial^2 \phi}{\partial x \partial y} \end{array}\right\}. \tag{10.2}$$

Using Equation (10.2) in Equation (10.1), we have

$$\begin{bmatrix} D_1 \left(\frac{\partial^4 \phi}{\partial x^4} + 2\frac{\partial^4 \phi}{\partial x^2 \partial y^2} + \frac{\partial^4 \phi}{\partial y^4} + \frac{\partial^2 \nu}{\partial x^2}\frac{\partial^2 \phi}{\partial y^2}\right) + \frac{\partial^2 D_1}{\partial x^2}\left(\frac{\partial^2 \phi}{\partial x^2} + \nu \frac{\partial^2 \phi}{\partial y^2}\right) \\[2mm] +2\frac{\partial D_1}{\partial x}\left(\frac{\partial^3 \phi}{\partial x^3} + \frac{\partial^3 \phi}{\partial x \partial y^2} + \frac{\partial \nu}{\partial x}\frac{\partial^2 \phi}{\partial y^2}\right) + \frac{\partial^2 D_1}{\partial y^2}\left(\frac{\partial^2 \phi}{\partial y^2} + \nu \frac{\partial^2 \phi}{\partial x^2}\right) \\[2mm] +2\frac{\partial D_1}{\partial y}\left(\frac{\partial^3 \phi}{\partial y^3} + \frac{\partial^3 \phi}{\partial y \partial x^2} - \frac{\partial \nu}{\partial x}\frac{\partial^2 \phi}{\partial x \partial y}\right) + 2(1 - \nu)\frac{\partial^2 D_1}{\partial x \partial y}\frac{\partial^2 \phi}{\partial x \partial y} \end{bmatrix}$$

$$+\rho l \frac{\partial^2 \phi}{\partial t^2} = 0. \tag{10.3}$$

The deflection of the plate can be written as a product of deflection function and time function as:

$$\phi(x, y, t) = \Phi(x, y) T(t). \tag{10.4}$$

Using Equation (10.4) in Equation (10.3), we have

$$T\begin{bmatrix} D_1 \left(\frac{\partial^4 \Phi}{\partial x^4} + 2\frac{\partial^4 \Phi}{\partial x^2 \partial y^2} + \frac{\partial^4 \Phi}{\partial y^4} + \frac{\partial^2 \nu}{\partial x^2}\frac{\partial^2 \Phi}{\partial y^2}\right) + \frac{\partial^2 D_1}{\partial x^2}\left(\frac{\partial^2 \Phi}{\partial x^2} + \nu \frac{\partial^2 \Phi}{\partial y^2}\right) \\[2mm] +2\frac{\partial D_1}{\partial x}\left(\frac{\partial^3 \Phi}{\partial x^3} + \frac{\partial^3 \Phi}{\partial x \partial y^2} + \frac{\partial \nu}{\partial x}\frac{\partial^2 \Phi}{\partial y^2}\right) + \frac{\partial^2 D_1}{\partial y^2}\left(\frac{\partial^2 \Phi}{\partial y^2} + \nu \frac{\partial^2 \Phi}{\partial x^2}\right) \\[2mm] +2\frac{\partial D_1}{\partial y}\left(\frac{\partial^3 \Phi}{\partial y^3} + \frac{\partial^3 \Phi}{\partial y \partial x^2} - \frac{\partial \nu}{\partial x}\frac{\partial^2 \Phi}{\partial x \partial y}\right) + 2(1 - \nu)\frac{\partial^2 D_1}{\partial x \partial y}\frac{\partial^2 \Phi}{\partial x \partial y} \end{bmatrix}$$

$$+\rho l \Phi \frac{\partial^2 T}{\partial t^2} = 0. \tag{10.5}$$

Using variable separable technique, we get

$$
\frac{\left[\begin{array}{l} D_1\left(\frac{\partial^4\Phi}{\partial x^4} + 2\frac{\partial^4\Phi}{\partial x^2\partial y^2} + \frac{\partial^4\Phi}{\partial y^4} + \frac{\partial^2\nu}{\partial x^2}\frac{\partial^2\Phi}{\partial y^2}\right) + \frac{\partial^2 D_1}{\partial x^2}\left(\frac{\partial^2\Phi}{\partial x^2} + \nu\frac{\partial^2\Phi}{\partial y^2}\right) \\[2mm] +2\frac{\partial D_1}{\partial x}\left(\frac{\partial^3\Phi}{\partial x^3} + \frac{\partial^3\Phi}{\partial x\partial y^2} + \frac{\partial\nu}{\partial x}\frac{\partial^2\Phi}{\partial y^2}\right) + \frac{\partial^2 D_1}{\partial y^2}\left(\frac{\partial^2\Phi}{\partial y^2} + \nu\frac{\partial^2\Phi}{\partial x^2}\right) \\[2mm] +2\frac{\partial D_1}{\partial y}\left(\frac{\partial^3\Phi}{\partial y^3} + \frac{\partial^3\Phi}{\partial y\partial x^2} - \frac{\partial\nu}{\partial x}\frac{\partial^2\Phi}{\partial x\partial y}\right) + 2\left(1 - \nu\right)\frac{\partial^2 D_1}{\partial x\partial y}\frac{\partial^2\Phi}{\partial x\partial y} \end{array}\right]}{\rho l\Phi}
$$

$$
= -\frac{1}{T}\frac{\partial^2 T}{\partial t^2} = \omega^2 \text{ (say).}
$$
(10.6)

By taking first and last expression of Equation (10.6), we get

$$
\left[\begin{array}{l} D_1\left(\frac{\partial^4\Phi}{\partial x^4} + 2\frac{\partial^4\Phi}{\partial x^2\partial y^2} + \frac{\partial^4\Phi}{\partial y^4} + \frac{\partial^2\nu}{\partial x^2}\frac{\partial^2\Phi}{\partial y^2}\right) + \frac{\partial^2 D_1}{\partial x^2}\left(\frac{\partial^2\Phi}{\partial x^2} + \nu\frac{\partial^2\Phi}{\partial y^2}\right) \\[2mm] +2\frac{\partial D_1}{\partial x}\left(\frac{\partial^3\Phi}{\partial x^3} + \frac{\partial^3\Phi}{\partial x\partial y^2} + \frac{\partial\nu}{\partial x}\frac{\partial^2\Phi}{\partial y^2}\right) + \frac{\partial^2 D_1}{\partial y^2}\left(\frac{\partial^2\Phi}{\partial y^2} + \nu\frac{\partial^2\Phi}{\partial x^2}\right) \\[2mm] +2\frac{\partial D_1}{\partial y}\left(\frac{\partial^3\Phi}{\partial y^3} + \frac{\partial^3\Phi}{\partial y\partial x^2} - \frac{\partial\nu}{\partial x}\frac{\partial^2\Phi}{\partial x\partial y}\right) + 2\left(1 - \nu\right)\frac{\partial^2 D_1}{\partial x\partial y}\frac{\partial^2\Phi}{\partial x\partial y} \end{array}\right]
$$

$$
-\rho\omega^2 l\Phi = 0.
$$
(10.7)

Equation (10.7) represents the equation motion for transverse vibration of the plate.

The rigidity of flux (of plate) is given by

$$
D_1 = \frac{El^3}{12\left(1 - \nu^2\right)}.
$$
(10.8)

10.4 Construction of Problem

Consider a non-homogeneous square plate of length a with one-dimensional linear variation in thickness l presented in Figure 10.1 as:

$$
l = l_0\left[1 + \beta\frac{x}{a}\right],
$$
(10.9)

where β, $(0 \leq \beta \leq 1)$ is a tapering parameter. The temperature distribution on the plate is considered to be linear in two dimensional as:

$$
\tau = \tau_0\left[1 - \left(\frac{x}{a}\right)\right]\left[1 - \left(\frac{y}{a}\right)\right],
$$
(10.10)

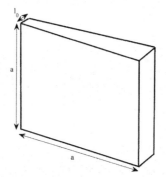

Figure 10.1 Square plate with linear variation in thickness.

where τ and τ_0 represents the temperature excess the reference temperature on the plate at any point and at the origin respectively. The modulus of elasticity is given by

$$E = E_0 \left(1 - \gamma\tau\right),\tag{10.11}$$

where E_0 is Young's modulus at $\tau = 0$ and γ is the slope of variation.

Using Equation (10.10) in Equation (10.11), we get

$$E = E_0 \left[1 - \alpha\left\{1 - \frac{x}{a}\right\}\left\{1 - \frac{y}{a}\right\}\right],\tag{10.12}$$

where α, $(0 \le \alpha < 1)$ is a temperature gradient, which is the product of temperature at origin and slope of variation i.e., $\alpha = \gamma\tau_0$. Since the plate is considered to be non-homogeneous, therefore the density (circular variation [31]) and Poisson's ratio (circular variation [20]) of the plate varies and are given by

$$\rho = \rho_0 \left[1 + m_1\left(1 - \sqrt{1 - \frac{x^2}{a^2}}\right)\right],\tag{10.13}$$

where ρ_0 is density at $x = 0$ and $m_1, (0 \le m_1 \le 1)$ is non-homogeneity constant and

$$\nu = \nu_0 \left[1 - m_2\left(1 - \sqrt{1 - \frac{x^2}{a^2}}\right)\right],\tag{10.14}$$

where ν_0 is Poisson's ratio at $x = 0$ and $m_2, (0 \le m_2 < 1)$ is another non-homogeneity constant.

Substituting Equations (10.9), (10.12) and (10.14) in Equation (10.8), we get

$$D_1 = \frac{E_0 l_0^3 \left[1 - \alpha \left\{1 - \frac{x}{a}\right\} \left\{1 - \frac{y}{a}\right\}\right] \left[1 + \beta \frac{x}{a}\right]^3}{12 \left[1 - \nu_0^2 \left\{1 - m_2 \left(1 - \sqrt{1 - \frac{x^2}{a^2}}\right)\right\}^2\right]} \tag{10.15}$$

10.5 Solution of Problem

In order to obtain time period, authors are using Rayleigh–Ritz technique (i.e., maximum kinetic energy T_s must equal to maximum strain energy V_s). Therefore, we must have

$$\delta \left(V_s - T_s\right) = 0, \tag{10.16}$$

where V_s and T_s are given by

$$V_s = \frac{1}{2} \int_0^a \int_0^a D_1 \left\{ \begin{matrix} \left(\frac{\partial^2 \Phi}{\partial x^2}\right)^2 + \left(\frac{\partial^2 \Phi}{\partial y^2}\right)^2 + 2\nu \frac{\partial^2 \Phi}{\partial x^2} \frac{\partial^2 \Phi}{\partial y^2} \\ +2\left(1 - \nu\right) \left(\frac{\partial^2 \Phi}{\partial x \partial y}\right)^2 \end{matrix} \right\} dy dx, \tag{10.17}$$

$$T_s = \frac{1}{2} \omega^2 \int_0^a \int_0^a \rho l \, \Phi^2 \, dy dx. \tag{10.18}$$

Authors compute the time period of frequency modes on clamped edges, therefore the boundary conditions are

$$\Phi = \frac{\partial \Phi}{\partial x} = 0, \text{ at } x = 0, \ a,$$

$$\Phi = \frac{\partial \Phi}{\partial y} = 0, \text{ at } y = 0, \ a. \tag{10.19}$$

The deflection function which satisfy Equation (10.19) is taken as

$$\Phi\left(x, y\right) = \left[\left(\frac{x}{a}\right) \left(\frac{y}{a}\right) \left(1 - \frac{x}{a}\right) \left(1 - \frac{y}{a}\right)\right]^2$$

$$\left[O_1 + O_2 \left(\frac{x}{a}\right) \left(\frac{y}{a}\right) \left(1 - \frac{x}{a}\right) \left(1 - \frac{y}{a}\right)\right], \tag{10.20}$$

where O_1 and O_2 are arbitrary constants. Substituting Equations (10.9), (10.13), (10.14) and (10.15) in Equations (10.17) and (10.18) we have

$$V_s = \frac{E_0 l_0^3}{24} \int_0^a \int_0^a \left[\begin{array}{c} \left\{ \dfrac{\left[1 - \alpha\{1 - \frac{x}{a}\}\{1 - \frac{y}{a}\}\right]\left[1 + \beta\frac{x}{a}\right]^3}{1 - \nu_0^2\left[1 - m_2\left(1 - \sqrt{1 - \frac{2}{a^2}}\right)\right]^2} \right\} \\ \left[\left(\frac{\partial^2 \Phi}{\partial x^2}\right)^2 + \left(\frac{\partial^2 \Phi}{\partial y^2}\right)^2 \right. \\ + 2\nu_0\left[1 - m_2\left(1 - \sqrt{1 - \frac{2}{a^2}}\right)\right]\frac{\partial^2 \Phi}{\partial\ ^2}\frac{\partial^2 \Phi}{\partial y^2} \\ \left. + 2\left(1 - \nu_0\left[1 - m_2\left(1 - \sqrt{1 - \frac{2}{a^2}}\right)\right]\right)\left(\frac{\partial^2 \Phi}{\partial\ \partial y}\right)^2 \right] \end{array} \right] dy\,dx,$$

(10.21)

$$T_s = \frac{1}{2}\omega^2 l_0\,\rho_0 \int_0^a \int_0^a \left[1 + m_1\left(1 - \sqrt{1 - \frac{x^2}{a^2}}\right)\right]\left[1 + \beta\frac{x}{a}\right]\Phi^2 dy\,dx.$$

(10.22)

Introducing non-dimensional variables X and Y as:

$$X = \frac{x}{a}, \quad Y = \frac{y}{a}.$$

(10.23)

Using Equation (10.23), Equations (10.21) and (10.22) converted into

$$V_s^* = \frac{E_0 l_0^3}{24a^2} \int_0^1 \int_0^1 \left[\begin{array}{c} \left\{ \dfrac{[1 - \alpha\{1 - X\}\{1 - Y\}][1 + \beta X]^3}{1 - \nu_0^2\left[1 - m_2\left(1 - \sqrt{1 - X^2}\right)\right]^2} \right\} \\ \left[\left(\frac{\partial^2 \Phi}{\partial X^2}\right)^2 + \left(\frac{\partial^2 \Phi}{\partial Y^2}\right)^2 \right. \\ + 2\nu_0\left[1 - m_2\left(1 - \sqrt{1 - X^2}\right)\right]\frac{\partial^2 \Phi}{\partial X^2}\frac{\partial^2 \Phi}{\partial Y^2} \\ \left. + 2\left(1 - \nu_0\left[1 - m_2\left(1 - \sqrt{1 - X^2}\right)\right]\right)\left(\frac{\partial^2 \Phi}{\partial X \partial Y}\right)^2 \right] \end{array} \right] dY\,dX,$$

(10.24)

$$T_s^* = \frac{1}{2}a^2\omega^2 l_0\,\rho_0 \int_0^1 \int_0^1 \left[1 + m_1\left(1 - \sqrt{1 - X^2}\right)\right]\left[1 + \beta X\right]\Phi^2 dY\,dX.$$

(10.25)

Using Equations (10.24) and (10.25), Equation (10.16) becomes

$$\delta\left(V_s^* - \lambda^2 T_s^*\right) = 0,$$

(10.26)

where $\lambda^2 = 12\rho_0\omega^2 a^4/E_0 l_0^2$ is frequency parameter. Equation (10.26) consists of two unknown constants O_1 and O_2. These two unknowns could be calculated as follows

$$\frac{\partial}{\partial\Omega_n}\left(V_s^* - \lambda^2 T_s^*\right) = 0, \quad n = 1, 2.$$

(10.27)

After simplifying Equation (10.27), we get system equations as:

$$c_{11}O_1 + c_{12}O_2 = 0,$$
$$c_{21}O_1 + c_{22}O_2 = 0. \tag{10.28}$$

To obtain frequency equation, the determinant of coefficient matrix of Equation (10.28) must zero i.e.,

$$\begin{vmatrix} c_{11} & c_{12} \\ c_{21} & c_{22} \end{vmatrix} = 0. \tag{10.29}$$

Equation (10.29) is quadratic equation from where we get frequency modes.

The time period of vibration of the plate is given by

$$K = \frac{2\pi}{\lambda}, \tag{10.30}$$

where λ is frequency modes obtained from Equation (10.29).

10.6 Numerical Illustration and Discussions

The time period of the frequency of the plate under temperature field, for different value of plate parameters (non-homogeneity constants m_1, m_2, thermal gradient α and tapered constant β) is calculated and presented with the help of tables. The following values of parameters are used for numerical calculations:

$$E_0 = 7.08 \times 10^{10} \text{N/m}^2, \rho_0 = 2.80 \times 10^3 \text{Kg/m}^3,$$
$$\nu_0 = 0.345, l_0 = 0.01\text{m}$$

Table 10.1 shows the time period with respect to non-homogeneity m_2 for a fixed value of another non-homogeneity constant m_1 i.e., $m_1 = 0.0$, for two different values of taper constant β i.e., $\beta = 0.0, 0.4$ and two different values of thermal gradient α i.e., $\alpha = 0.2, 0.6$. Table 10.1 provides the fact that with the increasing value of non-homogeneity constant m_2, the time period of frequency modes increase with less rate of increment for both the above-mentioned value of taper constant β and thermal gradient α. But time period of frequency modes decreasing with the increasing value of non-homogeneity m_2 when the tapered constant β varies from 0.0 to 0.4 and thermal gradient α varies from 0.2 to 0.6.

Table 10.1 Time period vs. non-homogeneity m_2 for $m_1 = 0.0$

m_2	$\beta = 0.0, \alpha = 0.2$		$\beta = 0.4, \alpha = 0.6$	
	K_1	K_2	K_1	K_2
0.0	0.042985	0.168224	0.036964	0.144707
0.2	0.043017	0.168314	0.037003	0.144740
0.4	0.043053	0.168359	0.037042	0.144773
0.6	0.043085	0.168450	0.037082	0.144807
0.8	0.043121	0.168495	0.037123	0.144840

Table 10.2 Time period vs. non-homogeneity m_2 for $m_1 = 0.4$

m_2	$\beta = 0.0, \alpha = 0.2$		$\beta = 0.4, \alpha = 0.6$	
	K_1	K_2	K_1	K_2
0.0	0.044460	0.173281	0.038305	0.149173
0.2	0.044495	0.173377	0.038347	0.149208
0.4	0.044530	0.173424	0.038387	0.149244
0.6	0.044564	0.173520	0.038429	0.149279
0.8	0.044599	0.173568	0.038469	0.149315

Table 10.2 displays second data set for the time period with respect to non-homogeneity constant m_2 for a fixed value of another non-homogeneity constant m_1 i.e., $m_1 = 0.4$, for two different values of taper constant β i.e., $\beta = 0.0, 0.4$ and two different values of thermal gradient α i.e., $\alpha = 0.2, 0.6$. Here also, the frequency modes increasing with increasing value of non-homogeneity m_2 and decreasing when tapered constant β and thermal gradient α varies from 0.0 to 0.4 and 0.2 to 0.6 respectively, like in Table 10.1. But the time period reported in Table 10.1 is less when compared with the time period reported in Table 10.2.

The time period for a fixed value of non-homogeneity m_2 i.e., $m_2 = 0.0$, for two different values of taper constant β i.e., $\beta = 0.0, 0.4$ and two different values of thermal gradient α i.e., $\alpha = 0.2, 0.6$ corresponding to a non-homogeneity constant m_1 is tabulated in Table 10.3. The increasing value of non-homogeneity m_1 results the increase in the time period for the above-mentioned value of taper constant β and thermal gradient α. When the combined value of taper constant β and thermal gradient α increases, the time period decreases. The rate of increment in the time period reported in Table 10.3 is high when compared with the rate of increment in the time period reported in Table 10.1. The time period reported in Table 10.1 (corresponding to Poisson's ratio) is less when compared with the time period reported in Table 10.3 (corresponding to density).

Table 10.4 provides the second data set of a time period corresponding to non-homogeneity constant m_1 for a fixed value of another non-homogeneity

Table 10.3   Time period vs. non-homogeneity m_1 for $m_2 = 0.0$

m_1	$\beta = 0.0, \alpha = 0.2$		$\beta = 0.4, \alpha = 0.6$	
	K_1	K_2	K_1	K_2
0.0	0.042985	0.168224	0.036964	0.144707
0.2	0.043727	0.170785	0.037641	0.146940
0.4	0.044460	0.173281	0.038305	0.149173
0.6	0.045176	0.175753	0.038955	0.151365
0.8	0.045886	0.178195	0.039599	0.153510

Table 10.4   Time period vs. non-homogeneity m_1 for $m_2 = 0.4$

m_1	$\beta = 0.0, \alpha = 0.2$		$\beta = 0.4, \alpha = 0.6$	
	K_1	K_2	K_1	K_2
0.0	0.043053	0.168359	0.037042	0.144773
0.2	0.043797	0.170924	0.037720	0.147009
0.4	0.044530	0.173424	0.038387	0.149244
0.6	0.045248	0.175901	0.039040	0.151438
0.8	0.045956	0.178347	0.039681	0.153585

Table 10.5   Time period vs. non-homogeneity m_1 and m_2

$m_1 = m_2$	$\beta = 0.0, \alpha = 0.2$		$\beta = 0.4, \alpha = 0.6$	
	K_1	K_2	K_1	K_2
0.0	0.042985	0.168224	0.036964	0.144707
0.2	0.043763	0.170831	0.037680	0.146975
0.4	0.044530	0.173424	0.038387	0.149244
0.6	0.045284	0.175999	0.039084	0.151475
0.8	0.046027	0.178499	0.039766	0.153660

m_2 i.e., $m_2 = 0.4$, for two different values of taper constant β i.e., $\beta = 0.0, 0.4$ and two different values of thermal gradient α i.e., $\alpha = 0.2, 0.6$. Like in Tables 10.1, 10.2 and 10.3, the time period is increasing with the increasing value of the non-homogeneity constant m_1 and decreases with combined increasing value of taper constant β and thermal gradient α. The time period reported in Table 10.4 is high when compared with the time period reported in Table 10.3.

Table 10.5 presents the time period corresponding to the simultaneous variation of non-homogeneity constants m_1, m_2 for two values of taper constant β i.e., $\beta = 0.0, 0.4$ and thermal gradient α i.e., $\alpha = 0.2, 0.6$. Here, the time period behaves the same as the time period reported in Tables 10.1, 10.2, 10.3 and 10.4. The simultaneous variation in both non-homogeneity constants m_1, m_2 (in Table 10.5) provides a high time period in comparison to a time period corresponding to variation in non-homogeneity constant m_1 (in Table 10.3) as well as variation in non-homogeneity constant m_2 (in Table 10.1).

Table 10.6 shows the time period of frequency modes corresponding to thermal gradient α by keeping non-homogeneity constant m_1 to be zero, for two different values of non-homogeneity constant m_2 and tapered constant β i.e., $m_2 = 0.2, 0.4$ and $\beta = 0.4, 0.6$. The increasing value of thermal gradient α results in the increasing in the time period of vibrational frequency, for the above-mentioned value of non-homogeneity constant m_2 and tapered constant β. On the other hand, the time period decreases when the non-homogeneity m_2 and taper constant β increases from 0.2 to 0.4 and 0.4 to 0.6 respectively.

Table 10.7 provides the second data set of a time period corresponding to thermal gradient α by taking non-homogeneity constant m_2 to be zero, for two different values of non-homogeneity constant m_1 and tapered constant β i.e., $m_1 = 0.2, 0.4$ and $\beta = 0.4, 0.6$. Here, the behavior of the time period is the same as reported in Table 10.6 in all manners. But the time period reported in Table 10.7 is high in comparison to the time period reported in Table 10.6.

The third data set of a time period corresponding to thermal gradient α for two different set values of non-homogeneity constants m_1, m_2 i.e., $m_1 = m_2 = 0.2, 0.4$ and tapered constant β i.e., $\beta = 0.4, 0.6$ is presented in Table 10.8. Like in Tables 10.6 and 10.7, the time period of frequency mode increases with the increasing value of thermal gradient and decreases when non-homogeneity constants m_1, m_2 and tapered constant β varies from 0.2 to 0.4 and 0.4 to 0.6 respectively. But the time period reported here, is higher than the time period tabulated in Tables 10.6 and 10.7.

Table 10.6 Time period vs. thermal gradient α for $m_1 = 0$

| | $m_2 = 0.2, \beta = 0.4$ | | $m_2 = 0.4, \beta = 0.6$ | |
α	K_1	K_2	K_1	K_2
0.0	0.034536	0.135064	0.031578	0.123393
0.2	0.035302	0.138061	0.032237	0.125991
0.4	0.036122	0.141290	0.032939	0.128727
0.6	0.037003	0.145074	0.033688	0.132949
0.8	0.037951	0.148433	0.034492	0.134803

Table 10.7 Time period vs. thermal gradient α for $m_2 = 0$

| | $m_1 = 0.2, \beta = 0.4$ | | $m_1 = 0.4, \beta = 0.6$ | |
α	K_1	K_2	K_1	K_2
0.0	0.035132	0.137157	0.032677	0.127215
0.2	0.035912	0.140187	0.033357	0.129871
0.4	0.036745	0.143451	0.034082	0.132696
0.6	0.037641	0.146940	0.034856	0.135735
0.8	0.038603	0.150748	0.035685	0.139008

Table 10.8   Time period vs. thermal gradient α

α	$m_1 = m_2 = 0.2, \beta = 0.4$		$m_1 = m_2 = 0.4, \beta = 0.6$	
	K_1	K_2	K_1	K_2
0.0	0.035168	0.137187	0.032747	0.127267
0.2	0.035949	0.140218	0.033430	0.129925
0.4	0.036784	0.143484	0.034158	0.132752
0.6	0.037680	0.146975	0.034937	0.135764
0.8	0.038622	0.150784	0.035748	0.139039

Table 10.9   Time period vs. taper constant β for $m_1 = 0$

β	$m_2 = \alpha = 0.4$		$m_2 = \alpha = 0.6$	
	K_1	K_2	K_1	K_2
0.0	0.044235	0.172662	0.045556	0.178044
0.2	0.039905	0.155987	0.041002	0.160162
0.4	0.036162	0.141322	0.037082	0.144807
0.6	0.032939	0.128729	0.033727	0.131695
0.8	0.030169	0.117927	0.030854	0.120482
1.0	0.027779	0.108611	0.028383	0.110873

Table 10.9 reflects the time period of frequency modes corresponding to taper constant β by taking non-homogeneity m_1 to be zero, for two different values of non-homogeneity m_2 and thermal gradient α i.e., $m_2 = \alpha = 0.4, 0.6$. The time period of frequency mode decreases with the increasing value taper constant for all the above-mentioned value of non-homogeneity m_2 and thermal gradient α. On the other hand, the time period of frequency mode increases as the combined value of non-homogeneity m_2 and thermal gradient α varies from 0.4 to 0.6. The rate of increment in a time period corresponding to the combined value of non-homogeneity m_2 and the thermal gradient α is very less.

Table 10.10 reflects the second data set of a time period corresponding to taper constant β by taking non-homogeneity m_2 to be zero, for two different values of non-homogeneity m_1 and thermal gradient α i.e., $m_1 = \alpha = 0.4, 0.6$. The behavior of the time period is the same as in Table 10.9 (in all respect). But the time period reported here is higher than the time period reported in Table 10.9.

Table 10.10   Time period vs. taper constant β for $m_2 = 0$

β	$m_1 = \alpha = 0.4$		$m_1 = \alpha = 0.6$	
	K_1	K_2	K_1	K_2
0.0	0.045679	0.178044	0.047762	0.185838
0.2	0.041241	0.160654	0.043035	0.167283
0.4	0.037395	0.145612	0.038955	0.151365
0.6	0.034082	0.132696	0.035462	0.137758
0.8	0.031231	0.121602	0.032462	0.126092
1.0	0.028770	0.112059	0.029880	0.116097

Table 10.11   Time period vs. taper constant β

β	$m_1 = m_2 = \alpha = 0.4$		$m_1 = m_2 = \alpha = 0.6$	
	K_1	K_2	K_1	K_2
0.0	0.045752	0.178195	0.047879	0.186003
0.2	0.041317	0.160777	0.043159	0.167462
0.4	0.037698	0.145680	0.039081	0.151475
0.6	0.034158	0.132752	0.035584	0.136709
0.8	0.031307	0.121649	0.032584	0.126143
1.0	0.028843	0.112079	0.029997	0.116097

The third data set of the time period with respect to taper constant β for two different values of non-homogeneity constants m_1, m_2 and thermal gradient α i.e., $m_1 = m_2 = \alpha = 0.4, 0.6$ is tabulated in Table 10.11. Time period decreases with the increasing value of tapering β and increases with the combined increasing value of non-homogeneity constants m_1, m_2 and thermal gradient α like in Tables 10.9 and 10.10. The time period reported here is also higher than the time period reported in Tables 10.9 and 10.10.

10.7 Results Comparison

A result comparison of the frequency of present analysis with [15] is presented in Table 10.12 with respect to non-homogeneity m_2 by keeping thermal gradient α to be zero for two different values of tapering constant β and non-homogeneity m_1 i.e., $\beta = m_1 = 0.0, 0.4$. It can be seen that frequency modes of the present study are less in comparison to frequency modes reported in [15]. The frequency modes of present analysis and [15] coincide at $m_2 = 0.0$.

Table 10.12   Non-homogeneity m_2 vs. frequency parameter (λ) for $\alpha = 0$

m_2	$m_1 = \beta = 0.0$		$m_1 = \beta = 0.4$	
	λ_1	λ_2	λ_1	λ_2
0.0	38.32	149.97	45.13	175.74
	{**38.32**}	{**149.97**}	{**45.13**}	{**175.74**}
0.2	38.30	149.86	45.12	175.57
	{**38.89**}	{**152.24**}	{**45.88**}	{**178.76**}
0.4	38.29	149.74	45.11	175.39
	{**39.67**}	{**155.34**}	{**46.91**}	{**182.99**}
0.6	38.27	149.63	45.09	175.22
	{**40.74**}	{**159.66**}	{**48.35**}	{**189.02**}
0.8	38.26	149.51	45.08	175.05
	{**42.23**}	{**165.84**}	{**50.40**}	{**197.82**}

The value in the bold bracket is from [15]

10.8 Conclusions

From the numerical illustration and comparison, the authors would like to record the following conclusions:

➤ The frequency modes due to circular variation in non-homogeneity m_2 (present study) provide less frequency when compared with the frequency modes due to exponential variation in non-homogeneity m_2 [15] as shown in Table 10.12.

➤ The time period corresponding to non-homogeneities m_1, m_2 and simultaneous variation in m_1, m_2 increases as shown in Tables 10.1, 10.2, 10.3, 10.4 and 10.5.

➤ Circular variation in m_2 (in Tables 10.1 and 10.2) provides less rate of increment in the time period in comparison to time period due to circular variation in m_1 (in Tables 10.3 and 10.4).

➤ The time period corresponding to m_2 is lesser than the time period corresponding to m_1 as shown in Tables 10.1 and 10.3.

➤ Time period with respect to m_2 is higher than the time period corresponding to m_1 when corresponding non-homogeneity constant varies from 0.0 to 0.2 and lesser when corresponding non-homogeneity constant varies from 0.6 to 0.8 (in Tables 10.2 and 10.4).

➤ The simultaneous variation of both non-homogeneity constants m_1, m_2 (in Table 10.5) provides a high time period in comparison to a time period corresponding to m_1 as well as corresponding to m_2.

➤ The time period increases when the temperature gradient increases as shown in Tables 10.6, 10.7, and 10.8.
➤ The time period decreases when the thickness of the plate increases as shown in Tables 10.9, 10.10, and 10.11.
➤ The time period corresponding to taper constant β and temperature α is higher when both the non-homogeneity constants m_1, m_2 include in the study in comparison to a time period where only either of non-homogeneity constants m_1 or m_2 includes.

Appendix: Nomenclature

a	Length of the plate
x, y	Coordinates in the plane of plate
M_x, M_y	Bending moment intensities in x and y direction
M_{xy}	Twisting moment intensity
E	Young's modulus
ν	Poisson's ratio
D_1	Flexural rigidity
ρ	Mass density per unit volume of the plate material
t	Time
$\phi(x, y, t)$	Deflection of plate
$\Phi(x, y)$	Deflection function
$T(t)$	Time function
l	Thickness of plate at x, y
β	Tapering parameter
m_1, m_2	Non-homogeneity constants
α	Temperature gradient

References

[1] S. Timoshenko, S. Woinowsky-Krieger, Theory of Plate and Shells, 1959.
[2] A. W. Leissa, Vibration of plates, Tech. Rep. sp-160, NASA, 1969.
[3] S. Chakarverty, Vibration of Plates, CRC Press, Taylor and Francis Group, Boca Raton, London New York, 2009.
[4] P. S. Timoshenko, J. M. Gere, Theory of Elastic Stability, McGraw-Hill New York, London, 2009.
[5] A. W. Leissa, Literature review: plate vibration research, 1976–1980: classical theory, The Shock and Vibration Digest, 13(9), 11–22, 1981.

[6] K. M. Liew, Y. Xiang, S. Kitipornchai, Research on thick plate vibration: a literature survey, Journal of Sound and Vibration, 180(1), 163–176, 1995.

[7] P. Paramasivam, Free vibration of square plates with square opening, Journal of Sound and Vibration, 30(2), 173–178, 1973.

[8] M. D. Olson, C. R. Hazell, Vibrations of a square plate with parabolically varying thickness, Journal of Sound and Vibration, 62(3), 399–410, 1979.

[9] B. Banerjee, Large amplitude vibrations of a clamped orthotropic square plate carrying a concentrated mass, Journal of Sound and Vibration, 82(3), 329–333, 1982.

[10] S. K. Chaudhuri, Large amplitude free vibrations of a square plate of variable thickness, Journal of Sound and Vibration, 92(1), 143–147, 1984.

[11] T. Sakiyama, M. Huang, Free vibrational analysis of rectangular plate with variable thickness, Journal of Sound and Vibration, 216(3), 379–397, 1998.

[12] W. L. Li, Vibration analysis of rectangular plate with general elastic boundary support, Journal of Vibration and Control, 273, 619–635, 2004.

[13] A. Sharma, V. Kumar, Vibrational frequency of isotropic square plate on C-C-S-S condition, Journal of Engineering and Applied Sciences, 12, 8719–8722, 2017.

[14] A. Sharma, V. Kumar, A. K. Raghav, Vibrational frequency of circular tapered square plate, Romanian Journal of Acoustics and Vibration, 14(1), 21–27, 2017.

[15] A. Sharma, Free vibration of square plate with temperature effect, Journal of Measurements in Engineering, 5(4), 222–228, 2017.

[16] A. Sharma, Vibration of square plate with parabolic temperature variation, Romanian Journal of Acoustics and Vibration, 14(2), 107–114, 2017.

[17] Y. F. Xiang, B. Liu, New exact solutions for free vibrations of thin orthotropic rectangular plates, Composite Structures, 89(4), 567–574, 2009.

[18] D. J. Gorman, A. Leissa, Free vibration analysis of rectangular plates, Journal of Applied mechanics, 49, 683, 1982.

[19] A. Sharma, Thermal induced vibration of non homogeneous tapered square plate, Journal of Engineering and Applied Sciences, 13, 2346–2351, 2018.

[20] A. Sharma, Vibration of skew plate with circular variation in thickness and poisson's ratio, Mechanics and Mechanical Engineering, 22(1), 43–52, 2018.

[21] A. Sharma, Natural vibration of parallelogram plate with circular variation in density, Acta Technica, 63(6), 763–774, 2018.

[22] M. A. Gharaibeh, A. M. Obeidat, Vibrations analysis of rectangular plates with clamped corners, Open Engineering, 8, 275–283, 2018.

[23] S. Sharma, R. Lal, N, Singh, Effect of non-homogeneity on asymmetric vibrations of non-uniform circular plates, Journal of Vibration and Control, 23(10), 1635–1644, 2017.

[24] Q. Wang, D. Shi, X. Shi, A modified solution for the free vibration analysis of moderately thick orthotropic rectangular plates with general boundary conditions, internal line supports and resting on elastic foundation, Meccanica, 51(8), 1985–2017, 2016.

[25] N. Szucs, G. Szeidl, Vibration of circular plates subjected to constant radial load in their plane, Journal of Computational and Applied Mechanics, 12(1), 57–76, 2017.

[26] W. Liu, D. Wang, H. Lu, Y. Cao, P. Zhang, Research on radial vibration of circular plate, Shock and Vibration, 2016, 8 pages, 2016.

[27] S. Soni, N. K. Jain, P. V. Joshi, Vibration analysis of partially cracked plate submerged in fluid, Journal of Sound and Vibration, 412, 28–57, 2018.

[28] K. Kalita, S. Haldar, Natural vibration of rectangular plate with-and without-rotary inertia, Journal of The Institution of Engineers (India): Series C, 539–555, 2018.

[29] A. Sharma, Vibration of nonhomogeneous square plate with circular variation in density, Engineering Vibration, Communication and Information Processing, Lecture Notes in Electrical Engineering, Book Series, 478, 253–263, 2019.

[30] A. Sharma, P. Kumar, Natural vibration of square plate with circular variation in thickness, Soft Computing: Theories and Applications, Part: Advances in Intelligent Systems and Computing, Book Series (AISC), 742, 311–319, 2019.

[31] A. Sharma, Vibration of visco-elastic plate with variable geometry, PhD Thesis, Amity University Haryana, 2018.

11

A Study on Metric Fixed Point Theorems Satisfying Integral Type Contractions

Naveen Mani[1,*], Amit Sharma[2], Reeta Bhardwaj[2] and Satishkumar M. Bhati[1]

[1]Department of Mathematics, Sandip University, Nashik, Maharashtra, India
[2]Department of Mathematics, Amity University Haryana, India
E-mail: naveenmani81@gmail.com; dba.amitsharma@gmail.com;
bhardwajreeta84@gmail.com; satishmbhati@gmail.com
*Corresponding Author

In the early 19th century, Cauchy proposed a method of successive approximation. These methods are helpful in finding the existence and uniqueness of solutions to a variety of models, particularly differential equations (ordinary as well as partial). A well renowned polish mathematician Banach [1] in 1922, developed and introduced a most prolific and applicable contraction principle. This principle is used to deduce the theorems on existence and uniqueness of solutions of various models. Branciari, in 2002, formulated and placed the notion of new contraction of integral type. Author proved a version of the contraction mapping principle satisfying a contraction of integral type in complete metric spaces. The main intent of this chapter is to derive some results on fixed point satisfying integral type weak contractions. An example and graphical representation of the fixed points of maps, and for weak contractions of our results are also given.

11.1 Introduction

This chapter contains a brief historical development of results, definitions, lemmas, and some theorems in complete metric spaces on a fixed point for maps concerning several integral types of contractive conditions. In the first section, we mention some basic backgrounds of results on fixed point satisfying contractions of integral types.

In section two, we derive some theorems satisfying rational weak contraction for real-valued functions of integral types. In support of our finding, an example with graphical representation has been given.

In section three, a result has been proven for a pair of mappings satisfying weak phi contraction of integral type. Our proved results are improved and extended version of some important results in the literature.

11.2 Preliminaries

Throughout the chapter, we assume that

$\Psi = \{\omega | \omega : [0, +\infty) \to [0, +\infty)\}$ is a Lebesgue integrable function which is summable on each compact subset of R^+, nonnegative, and is such that for each $\in > 0, \int_0^{\in} \omega(s)\, ds > 0$.

Fixed point operator theory has an escalating role in all branches of sciences (particularly in applied sciences, Mathematics and Physics). Accordingly, it becomes necessary to introduce this concept to the students at an early stage of their study. The first primary and most significant result of this theory was the well-known Banach-Caccioppoli theorem [1]. This theory was introduced in 1922. Afterward many escalating driving results have been derived and get focused in this modern era as well. One of the pioneer findings of this modern era is the integral type contraction introduced by Branciari [2] in 2002. Subsequently, many researchers extended the Branciari [2] result and derived some theorems on a fixed point. Some results satisfying the contractive condition of integral type are proved for single and multi-valued maps.

The author in [2] proved the following result.

Theorem 11.1 [2] *Let (L, d) be a complete metric space and $\lambda : L \to L$ be a map. If for every $k, l \in L$*

$$\int_0^{d(\lambda k, \lambda l)} \omega(s)ds \leq \rho \int_0^{d(k,l)} \omega(s)ds$$

where, $0 < \rho < 1$ and $\omega \in \Psi$, then there exists a $g \in L$ such that $\lambda g = g$.

This theorem was one of the real and proper extensions of the classical Banach result. The author in [2] by demonstrating some examples clarify the facts that it doesn't mean a function satisfies integral type contraction always has a fixed point, and always satisfies Banach contraction principle.

Example 11.1 [2] Let d be the Euclidean metric and $L = R_+$. Let us define a function $\lambda : [0, \infty) \to [0, \infty)$ as

$$\lambda(k) = k + 1 \quad and \quad w(s) = -1.$$

Then contraction in Theorem 1.1 is satisfied with all other conditions for any arbitrary k in $(0, 1)$. We can see that the map λ has no fixed point.

Following example does not satisfy Banach contraction but it satisfies the contraction given Theorem 1.1.

Example 11.2 [2] Let $L = \{\frac{1}{t} | t \in N\} \cup \{0\}$ with metric $d(u, v) = |u - v|$. Then (L, d) is complete. If we define a function $\lambda : L \to L$ as

$$\lambda(k) = \begin{cases} \frac{1}{t+1}; & if \ k = \frac{1}{t}, t \in N, \\ 0; & if \ k = 0, \end{cases}$$

then clearly contraction in Theorem 1.1 is satisfied on taking $w(s) = s^{\frac{1}{s-2}[1-\log s]}$ for $s > 0$, $w(0) = 0$, and $k = 0.5$. But the function defined above does not satisfy Banach result [1].

Rhoades [3], in 2003, generalized and extended the result of Branciari [2] by introducing some universal contraction.

Theorem 11.2 [3] *Let (L, d) be a complete metric space and $\lambda : L \to L$ be a map. If for every $k, l \in L$*

$$\int_0^{d(\lambda k, \lambda l)} w(s)ds \leq \rho \int_0^{R(k,l)} w(s)ds,$$

where,

$$R(k, l) = \max\left\{ d(k, l), d(k, \lambda k), d(l, \lambda l), \frac{d(k, \lambda l) + d(l, \lambda k)}{2} \right\},$$

$\rho \in [0, 1)$ *and* $w \in \Psi$, *then there exists a $g \in L$ such that $\lambda g = g$.*

Theorem 11.3 [3] *Let (L, d) be a complete metric space and $\lambda : L \to L$ be a map. If for every $k, l \in L$*

$$\int_0^{d(\lambda k, \lambda l)} \omega(s)ds \leq \rho \int_0^{R(k,l)} \omega(s)ds,$$

where,

$$R(k, l) = \max \left\{ d(k, l), d(k, \lambda k), d(l, \lambda l), d(k, \lambda l), d(l, \lambda k) \right\}$$

$\rho \in [0, 1)$ *and* $\omega \in \Psi$, *then there exists a* $g \in L$ *such that* $\lambda g = g$.

Kumar et al. [4], in 2007, proved an extended and generalized version of Branciari [2] by proving a result for compatible couple of maps satisfied a contraction of integral type. This result was step up as a version of Jungck's [15] theorem in metric spaces for compatible mappings.

Theorem 11.4 [4] *Let (L, d) be a complete metric space and λ, θ are compatible maps. If for every $u, q \in L$ following assumption satisfied:*

(i) $\lambda(L) \subset \theta(L)$, θ is continuous,

(ii) $\int_0^{d(\lambda u, \lambda q)} \omega(s)ds \leq \rho \int_0^{R(u,q)} \omega(s)ds$

$\rho \in [0, 1)$ *and* $\omega \in \Psi$, *then there exists a* $g \in L$ *such that* $\lambda g = g = \theta g$.

Example 11.3 [4] Let $L = \left\{ \frac{1}{t} : t \in N, t \neq 0 \right\} \cup \{0\}$ with the Euclidean metric d. Define mappingsλ, $\theta : L \to L$ by

$$\lambda(u) = \begin{cases} \frac{1}{t+4}, & \text{if } u = \frac{1}{t} \text{ and } t \text{ is odd,} \\ 0, & \text{other wise,} \end{cases}$$

and

$$\theta(u) = \begin{cases} \frac{1}{t+2}, & \text{if } u = \frac{1}{t} \text{ and } t \text{ is odd,} \\ 0, & \text{other wise.} \end{cases}$$

Let $\omega(s) = s^{\frac{1}{(s-2)}}(1 - \ln s)$, then hypothesis in Theorem 1.4 are satisfied. Also, $\lambda(0) = \theta(0) = 0$, is unique.

In 2008, Mocanu and Popa [5] derived few theorems on fixed point using iterations of sequences for maps in symmetric spaces under implicit relations.

Lemma 11.1 [5] *Let $\omega \in \Psi$ and sequence $\{d_t\}_{t \in N}$ be nonnegative with $\lim_{t \to \infty} d_t = g$ then*

$$\lim_{t \to \infty} \int_0^{d_t} \omega(s)ds = \int_0^g \omega(s)ds.$$

Lemma 11.2 [5] *Let* $\omega \in \Psi$ *and sequence* $\{d_t\}_{t \in N}$ *be nonnegative with* $\lim_{t \to \infty} d_t = g$ *then*

$$\lim_{t \to \infty} \int_0^{d_t} \omega(s)ds = 0 \quad \Rightarrow \quad \lim_{t \to \infty} d_t = 0.$$

In 2010, Samet and Yazidi [6] combined the result of Branciari [2] of the integral type with the result of Dass and Gupta [7] for rational expression and established a new theorem.

Theorem 11.5 [6] *Let* (L, d) *be a complete metric space and* $\lambda : L \to L$ *be a map. If for every* $k, l \in L$

$$\int_0^{d(\lambda u, \lambda q)} \omega(s)ds \leq \rho \int_0^{R(k,l)} \omega(s)ds + \vartheta \int_0^{d(k,l)} \omega(s)ds$$

and

$$R(k, l) = \frac{d(l, \lambda l)[1 + d(k, \lambda k)]}{[1 + d(k, l)]},$$

where $\rho, \vartheta > 0$ *are constants such that* $\rho + \vartheta < 1$ *and* $\omega \in \Psi$, *then there exists a* $g \in L$ *such that* $\lambda g = g$.

Liu et al. [8] extended and improved Branciari [2] result in sense of Kannan [9] contraction for real-valued function satisfying contractions of integral type. Liu et al. [8] deduced the following results:

Theorem 11.6 [8] *Let* (L, d) *be a complete metric space and* $\lambda : L \to L$ *be a map. If for every* $k, l \in L$

$$\int_0^{d(\lambda k, \lambda l)} \omega(s)ds \leq \tau(d(k, l)) \int_0^{d(k,l)} \omega(s)ds,$$

$\tau : R^+ \to [0, 1)$ *such that* $\limsup \tau(s) < 1$, $\forall\, s > 0$ *and* $\omega \in \Psi$, *then there exists a* $g \in L$ *such that* $\lambda g = g$.

Theorem 11.7 [8] *Let* (L, d) *be a complete metric space and* $\lambda : L \to L$ *be a map. If for every* $k, l \in L$

$$\int_0^{d(\lambda k, \lambda l)} \omega(s)ds \leq \tau_1(d(k, l)) \int_0^{d(k, \lambda k)} \omega(s)ds$$
$$+ \tau_2(d(k, l)) \int_0^{d(l, \lambda l)} \omega(s)ds,$$

$\tau_1, \tau_2 : R^+ \to [0,1)$ *with* $\tau_1(s)+\tau_2(s) < 1$, $\forall\, s \in R^+$, $\lim\limits_{s \to 0^+} \sup\, \tau_2(s) < 1$,

$\lim\limits_{s \to t^+} \sup\, \frac{\tau_1(s)}{1-\tau_2(s)} < 1$, $\forall\, s > 0$ *and* $\omega \in \Psi$, *then there exists a* $g \in L$ *such that* $\lambda g = g$.

From the past 15 years, many authors have continued their studies on integral type contractions and developed some fascinating results (see: [11–19]).

In the next two sections, we derive some results on fixed point satisfying integral type weak contractions.

11.3 Fixed Point Theorem Satisfying Weak Integral Type Rational Contractions

In this section, we derive fixed point theorem using the generalized rational expression for real-valued functions. Our results are an extension and improvement of Liu et al. [8] and Samet and Yazidi [6].

The main result of this paper is the following theorem.

Theorem 11.8 *Let f be a self-map and let (X, d) be a complete metric space. If for each $k \neq l \in X$*

$$\int_0^{d(fk,fl)} \omega(s)ds \le \gamma(d(k,l)) \int_0^{m(k,l)} \omega(s)ds \tag{11.1}$$

and

$$m(k,l) = \max\left\{ \frac{d(k,fk).d(l,fl)}{1+d(k,l)}, d(k,l) \right\}, \tag{11.2}$$

where $\omega \in \Psi, \gamma : R^+ \to [0,1)$ *is a function with*

$$\lim_{\delta \to t > 0} \sup \gamma(\delta) < 1. \tag{11.3}$$

Then f has a unique fixed point.

Proof. Let us consider $s \in X$ be any arbitrary point in X. Now construct a sequence $\{s_n\}$ in X such that $f s_n = s_{n+1}$.

Step - 1: Claim $\lim\limits_{n \to \infty} d(s_n, s_{n+1}) = 0$

For all $n \geq 0$, construct iteration on f by putting $k = s_n$ and $l = s_{n+1}$. From Equation (11.1), we get

$$\int_0^{d(s_n, s_{n+1})} w(s)ds = \int_0^{d(fs_{n-1}, fs_n)} w(s)ds$$
$$\leq \gamma(d(s_{n-1}, s_n)) \int_0^{m(s_{n-1}, s_n)} w(s)ds, \tag{11.4}$$

where

$$m(s_{n-1}, s_n) = \max\left\{ \frac{d(s_{n-1}, fs_{n-1})d(s_n, fs_n)}{1 + d(s_{n-1}, s_n)}, d(s_{n-1}, s_n) \right\}$$

$$= \max\left\{ \frac{d(s_{n-1}, s_n)d(s_n, s_{n+1})}{1 + d(s_{n-1}, s_n)}, d(s_{n-1}, s_n) \right\}$$

Since d is metric therefore for all n,

$$\frac{d(s_{n-1}, s_n)}{1 + d(s_{n-1}, s_n)} < 1 \text{ implies that } \frac{d(s_{n-1}, s_n)d(s_n, s_{n+1})}{1 + d(s_{n-1}, s_n)} < d(s_n, s_{n+1})$$

and hence

$$m(s_{n-1}, s_n) = \max\{d(s_n, s_{n+1}), d(s_{n-1}, s_n)\} \tag{11.5}$$

If we suppose that $d(s_n, s_{n+1}) > d(s_n, s_{n-1})$ then

$$m(s_{n-1}, s_n) = \max\{d(s_n, s_{n+1}), d(s_{n-1}, s_n)\} = d(s_n, s_{n+1})$$

On combining (11.4) & (11.3) implies

$$\int_0^{d(s_n, s_{n+1})} w(s)ds \leq \gamma(d(s_{n-1}, s_n)) \int_0^{d(s_n, s_{n+1})} w(s)ds < \int_0^{d(s_n, s_{n+1})} w(s)ds$$

This is not possible. Therefore $d(s_n, s_{n+1}) \leq d(s_n, s_{n-1})$ and hence from (11.5), $m(s_{n-1}, s_n) = d(s_{n-1}, s_n)$.
Therefore, from (11.4)

$$\int_0^{d(s_n, s_{n+1})} w(s)ds \leq \gamma(d(s_{n-1}, s_n)) \int_0^{d(s_{n-1}, s_n)} w(s)ds$$

$$\int_0^{d(s_n, s_{n+1})} w(s)ds \leq \int_0^{d(s_{n-1}, s_n)} w(s)ds \tag{11.6}$$

Thus, we obtained a monotone decreasing sequence $\left\{ \int_0^{d(s_n,s_{n+1})} w(s)ds \right\}$ of nonnegative real numbers and so we can find some $k \geq 0$ such that

$$\lim_{n \to \infty} \int_0^{d(s_n,s_{n+1})} w(s)ds = k \tag{11.7}$$

Suppose that $k > 0$, then on using (11.1) and (11.5) we have

$$0 < \int_0^k w(s)ds = \lim_{n \to \infty} \sup \int_0^{d(s_n,s_{n+1})} w(s)ds$$

$$\leq \lim_{n \to \infty} \sup \left(\gamma(d(s_{n-1}, s_n)) \int_0^{m(s_{n-1},s_n)} w(s)ds \right)$$

$$\leq \lim_{n \to \infty} \sup \gamma(d(s_{n-1}, s_n)) \lim_{n \to \infty} \sup \int_0^{\max\{d(s_n,s_{n+1}),d(s_{n-1},s_n)\}} w(s)ds$$

$$\leq \left(\lim_{s \to k} \sup \gamma(k) \right) \int_0^k w(s)ds < \int_0^k w(s)ds$$

This is not possible and so

$$\Rightarrow \lim_{n \to \infty} \int_0^{d(s_n,s_{n+1})} w(s)ds = 0$$

$$\Rightarrow \lim_{n \to \infty} d(s_n, s_{n+1}) = 0 \tag{11.8}$$

First, we claim that $\{s_n\}$ is a Cauchy sequence. Suppose it is not. That is there exist an $\varepsilon > 0$ and two sub sequence $\{s_{n_k}\}$, $\{s_{m_k}\}$ of $\{s_n\}$ the subject to $m_k > n_k \geq k$, $k > 0$ and satisfying

$$d(s_{m_k}, s_{n_k}) \geq \varepsilon \, \& \, d(s_{m_{k-1}}, s_{n_k}) < \varepsilon \tag{11.9}$$

For all $k = 0$, we have,

$$\int_0^\varepsilon w(s)ds \leq \int_0^{d(s_{m_k},s_{n_k})} w(s)ds \leq \gamma(d(s_{m_{k-1}}, s_{n_{k-1}})) \int_0^{m(s_{m_{k-1}},s_{n_{k-1}})} w(s)ds, \tag{11.10}$$

where

$$m(s_{m_{k-1}}, s_{n_{k-1}}) = \max \left\{ \frac{d(s_{m_{k-1}}, f s_{m_{k-1}}).d(s_{n_{k-1}}, f s_{n_{k-1}})}{1 + d(s_{m_{k-1}}, s_{n_{k-1}})}, d(s_{m_{k-1}}, s_{n_{k-1}}) \right\}$$

$$= \max \left\{ \frac{d(s_{m_{k-1}}, s_{m_k}).d(s_{n_{k-1}}, s_{n_k})}{1 + d(s_{m_{k-1}}, s_{n_{k-1}})}, d(s_{m_{k-1}}, s_{n_{k-1}}) \right\} \tag{11.11}$$

Triangle inequality implies

$$d(s_{m_{k-1}}, s_{n_{k-1}}) = d(s_{m_{k-1}}, s_{n_k}) + d(s_{n_k}, s_{n_{k-1}})$$

and so (11.9) gives

$$d(s_{m_{k-1}}, s_{n_{k-1}}) \leq \varepsilon + d(s_{n_k}, s_{n_{k-1}}).$$

Using (11.8) and taking $\lim\limits_{k \to \infty}$, we obtained,

$$\lim_{k \to \infty} d(s_{m_{k-1}}, s_{n_{k-1}}) = \varepsilon \tag{11.12}$$

Thus Equation (11.11) implies (using (11.9) & (11.12)), that

$$\lim_{k \to \infty} m(s_{m_{k-1}}, s_{n_{k-1}}) = \varepsilon \tag{11.13}$$

Taking $\lim\limits_{k \to \infty}$ in (11.10) and using (11.3), (11.12) and (11.13), we get

$$\int_0^\varepsilon \omega(s)ds \leq \lim_{k \to \infty} \sup \left(\int_0^{d(s_{m_k}, s_{n_k})} \omega(s)ds \right)$$

$$\leq \lim_{k \to \infty} \sup \left(\gamma(d(s_{m_{k-1}}, s_{n_{k-1}})) \int_0^{m(s_{m_{k-1}}, s_{n_{k-1}})} \omega(s)ds \right)$$

$$\leq \lim_{k \to \infty} \sup \left(\gamma(d(s_{m_{k-1}}, s_{n_{k-1}})) \right)$$

$$\lim_{k \to \infty} \sup \left(\int_0^{m(s_{m_{k-1}}, s_{n_{k-1}})} \omega(s)ds \right) < \int_0^\varepsilon \omega(s)ds$$

This gives a contradiction to our assumption. Thus, $\{s_n\}$ is a Cauchy sequence. Call the limit ν such that

$$\lim_{n \to \infty} f s_n = \nu. \tag{11.14}$$

Step - 3: Claim that $d(f\nu, \nu) = 0$.
Suppose not, that is, $d(f\nu, \nu) > 0$.
Hence

$$0 < \int_0^{d(f\nu, \nu)} \omega(s)ds = \int_0^{d(f\nu, f s_n)} \omega(s)ds \leq \gamma(d(\nu, s_n)) \int_0^{m(\nu, s_n)} \omega(s)ds, \tag{11.15}$$

where

$$m(v, s_n) = \max \left\{ \frac{d(v, fv)d(s_n, fs_n)}{1 + d(v, s_n)}, d(v, s_n) \right\}$$

Consequently, $\lim_{n \to \infty} m(v, s_n) = 0$

Hence on letting $\lim_{n \to \infty}$ Equation (11.15) implies

$$0 < \int_0^{d(fv,v)} w(s)ds = \int_0^{d(fv,fs_n)} w(s)ds \leq 0$$

This gives that $d(fv, v) = 0$. Therefore, $fv = v$.

Next, assume that there exists $\kappa \neq v$ s.t. $d(f\kappa, \kappa) = 0$.

Suppose $d(\kappa, v) \neq 0$.

Since,

$$0 < \int_0^{d(\kappa,v)} w(s)ds = \int_0^{d(f\kappa,fv)} w(s)ds \leq \gamma(d(\kappa, v)) \int_0^{m(\kappa,v)} w(s)ds \tag{11.16}$$

where

$$m(\kappa, v) = \max \left\{ \frac{d(\kappa, f\kappa).d(v, fv)}{1 + d(\kappa, v)}, d(\kappa, v) \right\} = 0,$$

Then (11.16) implies $0 < \int_0^{d(\kappa,v)} w(s)ds \leq 0$. This is a contradiction to the above fact that $d(\kappa, v) \neq 0$. Therefore, v is unique in such way that $fv = v$. Hence the result is established.

Following two results are consequence findings of our main result.

Corollary 11.1 *Let f be a self-map and let (X, d) be a complete metric space. If for each $k \neq l \in X$*

$$d(fk, fl) \leq \gamma(d(k, l))m(k, l)$$

and

$$m(k, l) = \max \left\{ \frac{d(k, fk).d(l, fl)}{1 + d(k, l)}, d(k, l) \right\},$$

$\gamma : R^+ \to [0, 1)$ *s.t.* $\lim_{\delta \to t} \sup \gamma(\delta) < 1 \; \forall \; t > 0$. *Then f has a unique fixed point.*

Corollary 11.2 *Let f be a self-map and let (X, d) be a complete metric space. If for each $k \neq l \in X$*

$$\int_0^{d(fk,fl)} w(s)ds \leq c \int_0^{m(k,l)} w(s)ds$$

and

$$m(k, l) = \max\left\{\frac{d(k, fk).d(l, fl)}{1 + d(k, l)}, d(k, l)\right\}$$

where $\omega \in \Psi$ and $c \in (0, 1)$ is a constant. Then f has a unique fixed point.

Proof. By taking $\alpha(t) = c$ for all $t \in R^+$ we get the result. $\square$

Theorem 11.9 *Let f be a self-map and let (X, d) be a complete metric space. If for each $k \neq l \in X$*

$$\int_0^{d(fk, fl)} \omega(s)ds \leq \alpha(d(k, l)) \int_0^{\lambda(k, l)} \omega(s)ds + \beta(d(k, l)) \int_0^{d(k, l)} \omega(s)ds \tag{11.17}$$

and

$$\lambda(k, l) = \frac{d(k, fk).d(l, fl)}{1 + d(k, l)}$$

where $\omega \in \Psi$ and $\alpha, \beta \in [0, 1)$ subjected to $\alpha(r) + \beta(r) < 1, \forall r > 0$

$$\lim_{s\to 0} \alpha(s) < 1, \lim_{s\to t} \frac{\beta(s)}{1 - \alpha(s)} < 1, \limsup_{s\to t} \alpha(s) < 1, \limsup_{s\to t} \beta(s) < 1 \tag{11.18}$$

Then there exists unique point v such that fv = v.

Proof. Choose $s \in X$ as any arbitrary point. Construct sequence $\{s_n\}$ such that $fs_n = s_{n+1}$. $\square$

For all $n \geq 0$, construct iteration on f by putting $k = s_n$ and $l = s_{n+1}$. Equation (11.17) gives

$$\int_0^{d(s_n, s_{n+1})} \omega(s)ds \leq \alpha(d(s_{n-1}, s_n)) \int_0^{\lambda(s_{n-1}, s_n)} \omega(s)ds + \beta(d(s_{n-1}, s_n)) \int_0^{d(s_{n-1}, s_n)} \omega(s)ds \tag{11.19}$$

where

$$\lambda(s_{n-1}, s_n) = \left\{\frac{d(s_{n-1}, fs_{n-1}).d(s_n, fs_n)}{1 + d(s_{n-1}, s_n)}\right\}$$
$$= \left\{\frac{d(s_{n-1}, s_n).d(s_n, s_{n+1})}{1 + d(s_{n-1}, s_n)}\right\}$$

Since d is metric therefore for all n,

$$\frac{d(s_{n-1}, s_n)}{1 + d(s_{n-1}, s_n)} < 1 \text{ implies that } \frac{d(s_{n-1}, s_n)d(s_n, s_{n+1})}{1 + d(s_{n-1}, s_n)} < d(s_n, s_{n+1})$$

and hence
$$\lambda(s_{n-1}, s_n) \leq d(s_n, s_{n+1}) \tag{11.20}$$

On making use of (11.20), Equation (11.18) in Equation (11.19) all together implies that

$$\int_0^{d(s_n,s_{n+1})} w(s)ds \leq \alpha(d(s_{n-1}, s_n)) \int_0^{d(s_n,s_{n+1})} w(s)ds$$
$$+ \beta(d(s_{n-1}, s_n)) \int_0^{d(s_{n-1},s_n)} w(s)ds$$

$$\int_0^{d(s_n,s_{n+1})} w(s)ds \leq \frac{\beta(d(s_{n-1},s_n))}{(1-\alpha(d(s_{n-1},s_n)))} \int_0^{d(s_{n-1},s_n)} w(s)ds$$
$$< \int_0^{d(s_{n-1},s_n)} w(s)ds \tag{11.21}$$

Ongoing this mode, we obtain

$$\int_0^{d(s_n,s_{n+1})} w(s)ds < \int_0^{d(s_{n-1},s_n)} w(s)ds < \dots < \int_0^{d(s_1,s_2)} w(s)ds < \int_0^{d(s_0,s_1)} w(s)ds$$

This implies that there exists a monotone sequence $\left\{ \int_0^{d(s_n,s_{n+1})} w(s)ds \right\}$ of non-increasing non-negative real numbers such that for every $k \geq 0$

$$\lim_{n\to\infty} \int_0^{d(s_n,s_{n+1})} w(s)ds = k \tag{11.22}$$

Suppose that $k > 0$. Thus, by going on in a similar way as in Theorem 1.8 we find that $k = 0$ and thus

$$\Rightarrow \lim_{n\to\infty} \int_0^{d(s_n,s_{n+1})} w(s)ds = 0$$
$$\Rightarrow \lim_{n\to\infty} d(s_n, s_{n+1}) = 0 \tag{11.23}$$

First, we claim that $\{s_n\}$ is Cauchy sequence. Suppose it is not. That is, there exist an $\varepsilon > 0$ and two sub sequence$\{s_{n_k}\}, \{s_{m_k}\}$ of $\{s_n\}$ the subject to $m_k > n_k \geq k, k > 0$ and satisfying

$$d(s_{m_k}, s_{n_k}) \geq \varepsilon \& d(s_{m_k-1}, s_{n_k}) < \varepsilon \tag{11.24}$$

Now for all $k \geq 0$, we have,

$$\int_0^\varepsilon w(s)ds \leq \int_0^{d(s_{m_k},s_{n_k})} w(s)ds$$

$$\leq \alpha(d(s_{m_k-1}, s_{n_k-1})) \int_0^{\lambda(s_{m_k-1},s_{n_k-1})} w(s)ds$$
$$+ \beta(d(s_{m_k-1}, s_{n_k-1})) \int_0^{d(s_{m_k-1},s_{n_k-1})} w(s)ds \tag{11.25}$$

Now,

$$\lambda(s_{m_{k-1}}, s_{n_{k-1}}) = \left\{ \frac{d(s_{m_{k-1}}, f s_{m_{k-1}}) \cdot d(s_{n_{k-1}}, f s_{n_{k-1}})}{1 + d(s_{m_{k-1}}, s_{n_{k-1}})} \right\}$$
$$= \left\{ \frac{d(s_{m_{k-1}}, s_{m_k}) \cdot d(s_{n_{k-1}}, s_{n_k})}{1 + d(s_{m_{k-1}}, s_{n_{k-1}})} \right\} \tag{11.26}$$

Taking $\lim\limits_{k \to \infty}$ in (11.26) and make use of (11.23), we obtain

$$\lim_{k \to \infty} \lambda(s_{m_{k-1}}, s_{n_{k-1}}) = 0. \tag{11.27}$$

Taking $\lim\limits_{k \to \infty}$ in (11.25) and using Equations (11.27), (11.12) and (11.18), we obtain

$$0 \le \int_0^\varepsilon \omega(s) ds \le 0 + \lim_{k \to \infty} \left[\sup \left(\beta(d(s_{m_{k-1}}, s_{n_{k-1}})) \right) \int_0^{d(s_{m_{k-1}}, s_{n_{k-1}})} \omega(s) ds \right]$$

$$\le \lim_{s \to t} \sup(\beta(s)) \lim_{k \to \infty} \sup \int_0^{d(s_{m_{k-1}}, s_{n_{k-1}})} \omega(s) ds) < \int_0^\varepsilon \omega(s) ds$$

which is a contradiction and hence our assumption is wrong. Therefore $\{s_n\}$ is a Cauchy sequence. Call the limit a such that

$$\lim_{n \to \infty} f s_n = a. \tag{11.28}$$

Claim that $d(fa, a) = 0$. Suppose it is not, i.e. $d(fa, a) \ne 0$. Consider,

$$0 < \int_0^{d(fa, a)} \omega(s) ds = \int_0^{d(fa, f s_n)} \omega(s) ds$$
$$\le \alpha(d(a, s_n)) \int_0^{\lambda(a, s_n)} \omega(s) ds + \beta(d(a, s_n)) \int_0^{d(a, s_n)} \omega(s) ds, \tag{11.29}$$

where

$$\lambda(a, s_n) = \frac{d(a, fa) \cdot d(s_n, f s_n)}{1 + d(a, s_n)}. \tag{11.30}$$

Taking $\lim\limits_{n \to \infty}$ in Equation (11.29), Equation (11.30) and make use of Equation (11.30) in Equation (11.29), we arrived at a contradiction. This proves $d(fa, a) = 0$. Therefore, $fa = a$. Uniqueness of f follows directly from Theorem 1.8. Hence, the result is established.

Theorem 11.10 *Let f be a self-map and let (X, d) be a complete metric space. If for each $k \neq l \in X$*

$$\int_0^{d(fk,fl)} w(s)ds \;\leq\; \gamma(d(k,l)) \left(\int_0^{\lambda(k,l)} w(s)ds \;+\; \int_0^{d(k,l)} w(s)ds \right)$$

and

$$\lambda(k,l) = \frac{d(k,fk)d(l,fl)}{1+d(k,l)}$$

$w \in \Psi$ *with* $\gamma : R^+ \to [0, 0.5)$ *in such way that,* $\lim\limits_{s \to t} \frac{\gamma(s)}{1-\gamma(s)} < 1, \forall\, t > 0.$

Then there exists unique point v such that fv = v.

Proof. Proof of the result is on the same line of Theorem 1.9. ☐

Example 11.4 Let $X = [0, 1]$. Define the metric $d(k,l) = |k - l| \,\forall\, k, l \in X$, which is complete with set X.
Let us define a rational map $f : X \to X$

$$f(k) = \frac{k}{k+1} \;;\quad \forall\;\; k \in X$$

Also $w \in \Psi$ is defined as $w(s) = 2s; \forall\, s \in R^+$.

Define $\gamma : [0, \infty] \to [0, 1]$ is defined by

$$\gamma(s) = \begin{cases} 0.5; & if\ \ s = 0 \\ \frac{1}{(1+s)}; & \forall\ s \in (0 + \infty) \end{cases}$$

Consider,

$$\int_0^{d(fk,fl)} w(s)ds = \int_0^{|fk-fl|} w(s)ds$$

$$= \frac{(k-1)^2}{((1+k)(1+l))^2}$$

$$\leq \frac{(k-1)^2}{((1+|k-l|))} \leq \gamma(d(k,l)) \int_0^{m(k,l)} w(s)ds$$

We can see that f (11) = 0. Figure 11.1(a) shows the fixed point of map f and Figure 11.1(b) shows that R.H.S. expression of Theorem 1.8 dominates the L.H.S. expression of Theorem 1.8 in $X = [0, 1)$ which validates our inequalities in Example 1.4.

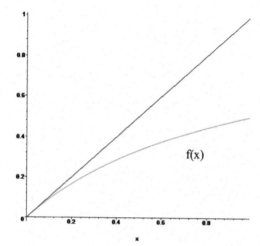

Figure 11.1(a)   Graph showing the fixed point of map f(x) defined in Example 1.4.

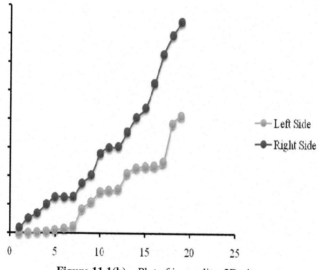

Figure 11.1(b)   Plot of inequality, 2D view.

11.4 $(\pi, \phi)-$ Integral Type Weak Contractions and Fixed-Point Theorems

Theorem 11.11 *Let S and T be compatible self-maps and (X, d) is a complete metric space, satisfying the following assumptions for all* $k, l \in X$

$$(i) \quad S(X) \subseteq T(X) \tag{11.31}$$

(ii) $\pi \left(\int\limits_0^{d(Sk,Sl)} \omega(s)ds \right) \leq \gamma(d(Tk,Tl))\phi \left(\int\limits_0^{d(Tk,Tl)} \omega(s)ds \right) , \quad (11.32)$

where $\omega \in \Psi, \pi : [0,+\infty) \to [0,+\infty)$ is continuous and non-decreasing with $\pi(t) = 0$ if and only if $t = 0, \phi : [0,+\infty) \to [0,+\infty)$ is a continuous map with the condition $\pi(t) > \phi(t)$ for all $t > 0$ and $\gamma : R^+ \to [0,1)$ is a function with

$$\lim_{\delta \to t > 0} \sup \gamma(\delta) < 1.$$

If T or S is continuous, then S and T have a unique common fixed point.

Proof. Choose x_0 as any arbitrary point of X. As $S(X) \subseteq T(X)$, so we can take a point x_1 in X in such a manner that $Sx_0 = Tx_1$. $\square$

Ongoing like this way, we can choose a sequence $\{x_n\}$ and define

$$y_n = Tx_{n+1} = Sx_n; \quad \forall \ n = 0, 1, 2, \ldots$$

For each integer $n \geq 1$ from (11.32), we obtain

$$\pi \left(\int\limits_0^{d(y_n,y_{n+1})} \omega(s)ds \right) \leq \gamma(d(y_{n-1}, y_n))\phi \left(\int\limits_0^{d(y_{n-1},y_n)} \omega(s)ds \right) \quad (11.33)$$

Using the conditions of Theorem 1.11, we get

$$\int\limits_0^{d(y_n,y_{n+1})} \omega(s)ds \leq \int\limits_0^{d(y_{n-1},y_n)} \omega(s)ds$$

Similarly,

$$\int\limits_0^{d(y_{n-1},y_n)} \omega(s)ds \leq \int\limits_0^{d(y_{n-2},y_{n-1})} \omega(s)ds$$

Above two inequalities imply that there exist a monotone sequence $\left\{ \int_0^{d(y_n,y_{n+1})} \omega(s)ds \right\}$ non increasing non-negative real numbers such that for every $k \geq 0$

$$\lim_{n \to \infty} \int_0^{d(y_n,y_{n+1})} \omega(s)ds = k \quad (11.34)$$

If on both sides of Equation (11.33), we take a limit as $n \to \infty$ and using (11.34) and conditions of Theorem 1.11, we obtain $\pi(k) \leq \phi(k)$. This is not possible for any real number. Therefore k=0.

This implies

$$\lim_{n\to\infty} \int_0^{d(y_n,y_{n+1})} w(s)ds = 0. \tag{11.35}$$

Now we prove that sequence $\{y_n\}$ is Cauchy sequence. Suppose on the contrary that it is not. Then there exists a $\varepsilon > 0$ with two subsequences $\{y_{n_k}\}$, $\{y_{m_k}\}$ of $\{y_n\}$ subjected to condition $k \le n_k < m_k$, $k > 0$ satisfying

$$d(y_{m_k}, y_{n_k}) \ge \varepsilon \text{ and } d(y_{m_{k-1}}, y_{n_k}) < \varepsilon \tag{11.36}$$

For all $k = 0$, we have

$$0 < H = \int_0^\varepsilon w(s)ds \le \int_0^{d(y_{m_k},y_{n_k})} w(s)ds \le \int_0^{\varepsilon+d(y_{n_k},y_{n_{k-1}})} w(s)ds$$

So from (11.35), we get

$$\lim_{k\to\infty} \int_0^{d(y_{m_k},y_{n_k})} w(s)ds = H \tag{11.37}$$

Also from triangle inequality, we know

$$d(y_{m_{k-1}}, y_{n_{k-1}}) = d(y_{m_{k-1}}, y_{n_k}) + d(y_{n_k}, y_{n_{k-1}}),$$

and therefore

$$\lim_{k\to\infty} \int_0^{d(y_{m_k}-1,y_{n_k}-1)} w(s)ds = H \tag{11.38}$$

From (11.32), we have

$$\pi\left(\int_0^{d(y_{m_k},y_{n_k})} w(s)ds\right) \le \gamma\left(d(y_{m_k}-1, y_{n_k}-1)\right) \phi\left(\int_0^{d(y_{m_k}-1,y_{n_k}-1)} w(s)ds\right)$$

Taking limit as $\lim_{k\to\infty}$ and using (11.37) and (11.38) along with the fact that π and ϕ are continuous, we get $\pi(H) \le \phi(H)$. This is a contradiction to the condition of Theorem 1.11. Hence $\{y_n\}$ is a Cauchy sequence. Therefore, there exists a point v such that

$$\lim_{n\to\infty} Tx_{n+1} = v \qquad \lim_{n\to\infty} Sx_n = v$$

Assume that T is continuous, therefore $v = \lim_{n\to\infty} Tx_{n+1} = T \lim_{n\to\infty} x_{n+1} = Tv$.

This proves that v is a fixed point of T.

Consider,

$$\pi \left(\int_0^{d(Sx_n,Sv)} \omega(s)ds \right) \leq \gamma(d(Tx_n,Tv))\phi \left(\int_0^{d(Tx_n,Tv)} \omega(s)ds \right)$$

On taking the limit as $n \to \infty$ and we get that $\int_0^{d(Sx_n,Sv)} \omega(s)ds = 0$ implies that $Sv = v$. Hence v is the point of coincidence of S and T. In Similar manner, we can deduce that coincidence point is unique. This proves the result.

11.5 Conclusions

From the above proved results, we draw the following conclusions.

1. From Corollary 2.1, we observed that the result proved in Theorem 1.8 is a genuine extension of the theorem proved by Rakotch [10].
2. From Corollary 2.2, we observed that result obtained in Theorem 1.8 is a proper extension of the theorem proved by Branciari [2].
3. Theorems 1.8 and Theorem 1.9 are proper extensions of the results of Liu et al. [8].
4. Theorem 1.11 is another extension of Branciari [2] for a pair of maps satisfying $(\pi, \phi)-$ integral type contraction.

References

[1] S. Banach, Sur les operations dans les ensembles abstraits et leur application aux equations integrals, Fundamenta Mathematicae. 1922, 3, 133–181.

[2] A. Branciari, A fixed point theorem for mappings satisfying a general contractive condition of integral type. International Journal of Mathematics and Mathematical Sciences. 2002; 29:531–536.

[3] BE Rhoades, Two fixed point theorems for mapping satisfying a general contractive condition of integral type. International Journal of Mathematics and Mathematical Sciences. 2003; 63:4007–4013.

[4] S. Kumar, R. Chugh, R. Kumar, Fixed point theorem for compatible mapping satisfying a contractive condition of integral. Soochow Journal of Mathematics. 2007; 33(11):181–185.

[5] M. Mocanu and V. Popa, Some fixed point theorems for mapping satisfying implicit relations in symmetric spaces. Libertas Math. 2008; 28:1–13.

[6] B. Samet and H. Yazidi, Fixed point theorems with respect to a contractive condition of integral type. Rend. Cric. Mat. Palermo. 2011; 60:181–190.

[7] DS. Jaggi, Some unique fixed point theorems, Indian Journal of Pure and Applied Mathematics, 1977, 8, 223–230.

[8] ZQ. Liu, Li X, SM Kang, SY Cho, Fixed point theorems for mappings satisfying contractive conditions of integral type and applications. Fixed Point Theory and Applications. 2011; 2011: Article Id 64.

[9] R. Kannan, Some results on fixed points, Bulletin of Calcutta Mathematical Society, 1968, 60, 71–76.

[10] E. Rakotch, A note on contractive mappings, Proceedings of the American Mathematical Society. 1962, 13, 459–465.

[11] NV. Luong and NX. Thuan, A fixed point theorem for $\varphi \int_\phi$ – weakly contractive mapping in metric spaces. Int. Journal of Math. Analysis. 2010; 4:233–242.

[12] ZQ Liu, SM. Li J,Kang, Fixed point theorems of contractive mappings of integral type. Fixed Point Theory and Applications. 2013; 2013: Article Id 300.

[13] ZQ Liu, SM. Li J,Kang, Ume JS. Fixed point theorems of contractive mappings of integral type. Fixed Point Theory and Applications. 2014; 2014:Article Id 394.

[14] G. Jungck, Compatible mappings and common fixed points, International Journal of Mathematics and Mathematical Sciences, 1986; 9 771–779.

[15] N. Mani, Existence of fixed points and their applications in certain spaces [Ph.D.Thesis], M. M. University, Mullana, Ambala, India, (2017).

[16] N. Mani, Generalized C_β^ψ – rational contraction and fixed point theorem with application to second order differential equation, Mathematica Moravica, 2018; 22(1), 43–54.

[17] G. Jungck, Compatible mappings and common fixed points, International Journal of Mathematics and Mathematical Sciences, 1988; 11, 285–288.

[18] V. Gupta and N. Mani, A common fixed point theorem for two weakly compatible mapping satisfying a new contractive condition of integral type, Mathematical Theory and Modeling, 2011; 1(1), pp 1–6.

[19] V. Gupta, N. Mani, Arslan H. Ansari, Generalized integral type contraction and common fixed point theorems using an auxiliary function, Advances in Mathematical Sciences and Applications, 2018; 27(2), 263–275.

[20] V. Gupta and N. Mani, A fixed point theorem satisfying a generalized weak contractive condition of integral type, International Journal of Mathematical Analysis, 2012; 6(38), 1883–1889.

12

Objective Function – in Radiometric Studies – Application to AGRS Surveys Associated with Radon

Shivvaran Singh Raghuwanshi[1,2]

[1]Ex-Head Physics Group, Atomic Minerals Directorate for Exploration & Research, Department of Atomic Energy, Hyderabad, India
[2]Bhopal (M.P.), India
E-mail: raghu2257@yahoo.com

Variation of stripping factors observed during the study of field test strip data obtained by low altitude gamma ray spectrometric surveys and calibration test pad have been analyzed by ordinary and weighted least square models. Mathematical analyses of data were carried out when both correlatable and non-correlatable conditions exist in between different radioelements. The effects of airborne Bi^{214} and its error bias on the results have been taken care of. Study reveals that the variation of stripping factors can also be achieved even when correlation between radioelements exists in the nature.

12.1 Introduction

Various studies on the stripping factors in airborne gamma ray spectrometry have been done at various times (Dickson et al., 1981, Grasty, 1975, Loveborg, 1978, Jadhav et al. 1988, Raghuwanshi, 1987, 1989, 1992). The airborne data for Uranium (U), Thorium (Th), Potassium (K) is being collected in respective energy windows by airborne gamma ray spectrometric surveys over a field test strip where ground concentration of radioelements is known. The flying altitude varies from 30 m to 250 m. This process is repeated as routine exercise before the commencement and at the end of airborne gamma-ray spectrometric (AGRS) surveys by Atomic Minerals Directorate for Exploration and Research (AMD) for exploration of atomic minerals. Auxiliary data like temperature and pressure along with radioelements concentration are also recorded for determining the atmospheric conditions which affect the concentration of radio-elements. Apart from test strip, calibration pad (Jadav et al., 1988) data is also obtained for calibration of data acquisition system.

In the present work an objective function is derived for the study of the variation of the stripping factors which plays a significant role in knowing the error behavior of the Compton scattering ratios in Airborne gamma ray spectrometric investigations. Subsequently, statistical methods are adopted to analyze the airborne radiometric data using test areas in varied conditions and probable models of ordinary and weighted least squares are used and the results have been compared and are discussed in this paper.

12.2 Ordinary and Weighted Linear Regression Models

Let K, U, Th be the counts rates observed in AGRS system on either calibration pads or on the test strips. Let B_k, B_U, B_{Th} be the background counts in the respective energy channels of potassium (1.36–1.46 MeV), uranium (1.66–1.86 MeV) and thorium (2.42–2.82 MeV). Then we can write the count equations as follow:

$$
\begin{aligned}
K &= S_{uk}K_c + S_{ku}U_c + S_{kt}Th_c + B_k \\
U &= S_{uk}K_c + S_{uu}U_c + S_{ut}Th_c + B_u \\
Th &= S_{Tk}K_C + S_{tu}U_C + S_{tt}Th_C + B_{Th}
\end{aligned}
\tag{12.1}
$$

Where S_{ij}'s are sensitivities to be determined and give count rates in ith channel/window per unit concentration of jth radioelement.

The solution of the above calibration equations is sought by the use of the least square principle using calibration pads data where concentration of U, Th, K are known. In this approach, the total error between the observed and predicted counts is minimized with respect to the parameters to be optimally computed. viz., sensitivities, interchannel coefficients. This can be done in two ways: One considers no uncertainties in the count rates and solution of calibration constants are found using ordinary least squares (OLS) method. While in the other method due weightage to uncertainties in the variables (count rates) is given. And the techniques applied is called the weighted least squares technique (WLS). These two ways shall be studied in what follows.

These two models are discussed as follows using a linear relationship between two variables x and y:

$$y_i = a + bx_i + E_i \tag{12.2}$$

where $E_i's$ are errors due to misfit of the line $y = a + bx$ to experimental data. In this expression y_i and x_i are priori known for ith point (x_i, y_i) and a and b are parameters to be estimated by minimizing total E_i^2 with respect to a and b. If the variances associated with y_i, x_i be $\sigma^2 y_i$, $\sigma^2 x_i$ respectively, then the weight w_i is defined as

$$w_i = \frac{1}{\sigma^2 y_i + b^2 \sigma^2 x_i} \tag{12.3}$$

In this method an initial guess to 'b' is made. Then a series of regressions are performed till b does not show successively any further appreciable change (by setting a tolerance).

These concepts provide the following two sets of parameters a and b (a) OLS mode.

$$b = \frac{\sum_{i=0}^{n} (x_i - x)(y_i - y)}{\sum_{i=0}^{n} (x_i - x)^2} \tag{12.4}$$

And

$$a = y - bx \tag{12.5}$$

where
(a)

$$x = \frac{\sum_{i=0}^{n} (x_i)}{n}$$

and

$$y = \frac{\sum_{i=0}^{n} (y_i)}{n}$$

(b) in WLS mode:

$$x = \frac{\sum_{i=0}^{n} (w_i x_i)}{\sum_{i=0}^{n} (w_i)}$$

$$y = \frac{\sum_{i=0}^{n} (w_i y_i)}{\sum_{i=0}^{n} (w_i)}$$

and first estimate of b is guessed value given by expression (12.4) to start with.

OLS is usually regarded as a mathematical method of fitting the best straight line to the (x_i, y_i) points but it is not strictly so. Because an assumption used in deriving normal equations of regression is that the dependent variable y_i contains random error of measurement, while independent variable (x_i) is free from random error and other assumption is that the absolute error of measurement in y values remains constant over the whole range of values of x_i.

The WLS is an extension of OLS and in it the observations are given due weightage by a factor inversely proportional to the variance of x_i. The variance in radiometric data is square root of the count rates because the radiometric counts follow a Poisson distribution. In this approach of weighing, the regression gives less importance to the data with large variances and the line of best fit passes near those points with less variances. It, therefore, provides a more realistic fit to the radiometric data sensitive to errors.

12.3 Computations of Compton Factors

Let the observed counts in the uranium, thorium and potassium windows be U_G, Th_G, K_G, and the background measurements be U_{BG}, T_{BG} and K_{BG} and U_N, Th_N, and K_N be the net counts in these energy windows respectively. Then following relations can be written:

$$U_N = U_G - \alpha Th_N - U_{BG} \tag{12.6a}$$

$$Th_N = Th_G - T_{BG} \tag{12.6b}$$

$$K_N = K_G - \beta Th_N - \gamma U_N - K_{BG} \tag{12.6c}$$

Where α, β are the Compton scattering contributions of thorium gamma rays (2.62 MeV) in the uranium and potassium windows respectively and γ is the Compton scattering contribution of uranium gamma rays in the potassium window. These are called stripping factors and are caused due to scattering of gamma rays in the intervening medium. In airborne surveys therefore, they vary with altitude of flying. These stripping factors at various altitudes help in computing net counts of radioelements. (Grasty, 1975; IAEA, 1991). In practice, these values are estimated on calibration pads with specific requirements (Lovberg, 1984). However, the variation of these stripping factor with altitude is either estimated by simulation (Dickson et al., 1981) or by Monte-Carlo numerical experiments (Lovberg, et al., 1978). Raghuwanshi et al. (1987) were first to have directly determined these coefficients.

Now if the ground distribution of uranium and potassium concentration be fairly constant and the concentration of the thorium varies significantly compared to the uranium and the potassium concentrations, it can be shown that these conditions are sufficient to estimate the stripping factors using the airborne data for U_G, Th_G and K_G on this kind of ground by actual flying over such a ground by the instrumented aircraft. Let us put the above equations in the following form:

$$U_G = \alpha \, Th_G + A_1(h) \tag{12.7a}$$

$$K_G = (\beta(h) - \gamma(h))Th_G + \gamma(h)U_G + A_2(h) \tag{12.7b}$$

Where $A_1(h)$ and $A_2(h)$ are altitude dependent constants; and $A_1(h)$ depends upon U_N, U_{BG}; and $A_2(h)$ on T_{BG}, $\alpha(h)$, $\beta(h)$, $\gamma(h)$, K_{BG} & U_{BG}, K_N. These factors $A_1(h)$ and $A_2(h)$ can be treated as independent of U_N and K_N, respectively by way of our above-mentioned assumption but will depend upon the altitude because of the varying α, β, γ's. For the range of airborne survey altitudes, the backgrounds T_{BG}, K_{BG}, U_{BG} (provided no air Bi^{214} is these or corrected for it) can be treated constant.

Therefore, Equations (12.7a) and (12.7b) form the basis for estimation of $\alpha(h)$ $\beta(h) - \gamma(h)$. Let $\beta(h) - \gamma(h) = T(h)$, say. From the knowledge of $T(h)$ one can get the value of $\beta(h)$ using α, γ values.

Therefore, from the airborne data for uranium and potassium acquired over an area or test strip with lower variation of uranium and potassium concentrations compared to the high variation of thorium concentration, it is possible to estimate the stripping factors from above equations provided there is no correlation among radioelements. By estimating them at various altitudes using Equation (12.7a) for $\alpha(h)$ and Equation (12.7b) for $T(h)$ & $\gamma(h)$, their altitude dependence can be computed optimally.

12.3.1 Study of Variations in Stripping Factors when the Correlation among Radioelements Exists

In the previous section we mentioned that Equations (12.7a) and (12.7b) can be used for estimation of the stripping factors $\alpha(h)$, $T(h)$ provided there were no correlations among K_N, U_N, Th_N. However, in the following we shall prove that the variation of the stripping factors with altitude is possible even in the case there exists a correlation among the radioelements.

Let us assume that the potassium counts K_N bear a definite relation with U_N and Th_N mentioned below:

$$K_N = mTh_N + qU_N + K_{N0} \tag{12.8}$$

Where m is partial correlation coefficient between K_N and Th_N; and q is that between K_N and U_N respectively. K_{N0} is a constant term and can be interpreted as counts available without the influence of Th_N and U_N. Let us subtract Equation (12.8) from Equation (12.6c), so as to eliminate K_N. After rearrangements of various terms, we can write it finally as

$$K_G = (m + \beta)Th_N + (\gamma + q)U_N + (K_{BG} + K_{N0}) \tag{12.9}$$

On using Equations (12.6a) and (12.6b), it can be rewritten as

$$K_G = [(\beta - \gamma\alpha) + (m - \alpha q)]Th_G + (\gamma + q)U_G \, Aa \tag{12.10}$$

Where $Aa = \alpha(\gamma + q)T_{BG} - (\gamma + q)U_{BG} - (m + \beta)T_{BG}$ has been substituted and can be treated as constant at an altitude and for the definite relationship (12.8).

Now let us write the coefficient of Th_G in Equation (12.10) as TT(h) at particular altitude h, i.e.

$$TT(h) = \beta(h) - \alpha(h)\,\gamma(h) + m - \alpha(h)q \tag{12.11}$$

Here all the stripping factors have been treated as the function of the height h.

In order to see the rates of changes of these stripping factors with the height we can differentiate TT(h) with respect to h. Since m and q are computable from the ground concentration of the radioelements and can be directly used to compute the variations of α, β, γ with altitude. In case potassium bears no correlation with uranium then q will be zero and rate of change of β can be estimated. We can conclude, therefore, that for the study of variation of stripping factors with altitude, we need only three necessary conditions:

(a) the uranium and potassium should not have the correlation;
(b) the uranium and thorium may or may not have the correlation;
(c) the variation of thorium concentration must be significantly more than the variation of U and K concentrations to make the data feasible for weighted or ordinary least squares regression.

In case the U_N and Th_N are correlated, then also the rate of change of stripping factor $\alpha(h)$ will remain unaffected.

12.3.2 Estimation of Stripping Factors and their Variations with Flying Heights

We shall now focus on the estimation of factors $\alpha(h)$, $\beta(h)$, $\gamma(h)$ given in Equation (12.7a) and (12.7b) over a range of altitudes by considering the data at every altitude separately. The estimation of $\alpha(h)$, $\beta(h)$, $\gamma(h)$ can be made by least squares procedure described earlier by minimizing the error with respect to these unknowns. Therefore, let the Equation (12.7b) be written as below (in case there is no correlation exists among U_N, Th_N, K_N)

$$K_{Gi} = (\beta(h) - \gamma(h))Th_{Gi} + \gamma(h)U_{Gi} + A_2(h) + \varepsilon_i \qquad (12.12)$$

where $\varepsilon_i = $ error, K_{Gi}, Th_{Gi}, U_{Gi} are observed counts for ith sample in potassium, thorium and uranium windows respectively. Putting

$$T(h) = \beta(h) \qquad (12.13)$$

Let us put for N number of samples, therefore, the total error (E) as follows:

$$E = \Sigma \; \varepsilon_i^2 \qquad (12.14)$$

Let the estimated values of $T(h)$ and $\gamma(h)$ be denoted by $\hat{T}(h)$ and $\hat{\gamma}(h)$ respectively. Then, the following two normal equations for determining $T(h)$ and $\gamma(h)$ can written as

$$\frac{\partial E}{\partial T(h)} = 0 = (S_{13} - S_{10}) - T(h)S_{11} - \gamma(h)S_{12} \qquad (12.15)$$

$$\partial E/(\partial \gamma(h)) = 0 = (S_{23} - S_{20}) - T(h)S_{21} - \gamma(h)S_{22} \qquad (12.16)$$

where

$$
\begin{aligned}
S_{13} &= \sum_{i=0}^{n}(K_{Gi} - Th_{Gi}), & S_{10} &= \sum_{i=0}^{n}(A_2 Th_{Gi}) \\
S_{23} &= \sum_{i=0}^{n}(U_{Gi} - K_{Gi}), & S_{20} &= \sum_{i=0}^{n}(A_2 U_{Gi}) \\
S_{11} &= \sum_{i=0}^{n}(Th_{Gi} Th_{Gi}), & S_{21} &= \sum_{i=0}^{n}(Th_{Gi} U_{Gi}) \\
S_{12} &= \sum_{i=0}^{n}(Th_{Gi} U_{Gi}), & S_{22} &= \sum_{i=0}^{n}(U_{Gi} U_{Gi})
\end{aligned}
\qquad (12.17)
$$

Solving the two Equations (12.15) and (12.16) for $T(h)$ and $\gamma(h)$ will provide us with necessary estimated values of these stripping factor provided the determinant

$$\Delta = \begin{vmatrix} S_{11} & S_{12} \\ S_{21} & S_{22} \end{vmatrix}$$

is not zero, then we can write

$$T(h) = \frac{(S_{13}-S_{10})S_{22}-(S_{23}-S_{20})S_{12}}{\triangle} \tag{12.18}$$

$$\gamma(h) = \frac{(S_{23}-S_{20})S11-(S_{13}-S_{10})S_{21}}{\triangle} \tag{12.19}$$

If in the Equations (12.18) and (12.19), $S_{11} \cdot S_{12} \simeq S_{12} \cdot S_{21}$ then the values of $T(h)$ and $\gamma(h)$ will become unstable i.e. in other words the counts in thorium and uranium windows will affect values of both $T(h)$ and $\gamma(h)$.

In the similar manner, the estimated value of $\alpha(h)$ can be written as follows.

$$\alpha(h) = \frac{S_{12} - S_{10}^{\alpha}}{S_{11}} \tag{12.20}$$

where

$$S_{10}^{\alpha} = \sum_{i=0}^{n} (A_1 - Th_{Gi})$$

After finding $\alpha(h)$, $T(h)$ and $\gamma(h)$, we can find $\beta(h)$ using relation (12.13). Then knowing these values at various altitudes h, their gradient with h can be determined in a similar manner by both least square procedures OLS and WLS.

12.4 Objective Function for Study of Random Errors in Stripping Coefficients

Let us find the confidence limits in the estimated values of the stripping coefficients $\alpha(h)$, $T(h)$ and $\gamma(h)$. We will describe this procedure in a general manner for $T(h)$. Let minimum of E be denoted by M_o and be expressed as follows:

$$M_0 = \sum_{i=0}^{n} \{(K_{Gi} - T(h)Th_{Gi} - \gamma(h)U_{Gi} - A_2)\}^2 \tag{12.21}$$

Let us call the predicted value K_{Gi}^P of K_{Gi} and is defined by

$$K_{Gi}^P = T(h)Th_{Gi} - \gamma(h)U_{Gi} - A_2 \tag{12.22}$$

then we can rewrite the minimum error as defined in Equation (12.21) as below:

$$M_0 = \sum_{i=0}^{n} \{(K_{Gi} - K_{Gi}^P)\}^2 \tag{12.23}$$

Since K^P_{Gi} is expressed through a linear relationship (12.22) of T(h) and γ(h), M_o has (N-2) degrees of freedom. Now let us find the sum of squares of the errors $\in_i$ for the values of T(h) and γ(h) using between the limits of one standard deviation i.e. 68% of confidence. Let the function M_o become M in such a case. Then the amount of total error will increase by $M-M_o$ due to above change in the value of T(h) and γ(h). Since we assume that $\in_i$ is normally distributed and M and M_o are the sum of their squares for N samples, we can say that M_o is K-squared distributed with (N-2) degrees of freedom, K^2_{N-2}. And $M-M_o$ is distributed like K^2_1 because $M-M_o$ has only one degree of freedom as we are only varying T(h) and keeping γ(h) constant in this process. Let us form a random variable by taking the ratio of two K^2 distributed variables. Then we can write

$$((M - M_0)/1)/(M/(N - 2))$$

a function. In such a case the function will be Fischer function F and we can write then

$$M - M_0 = \left(\frac{M_0}{N - 2}\right) F(1, N - 2, \; 68\%) \qquad (12.24)$$

This is the required *objective function* for determining the confidence value of the variable T(h) for 68% confidence. When the number of samples N tends to very large, the objective function $M-M_o$ can be written with sufficient accuracy as follows:

$$M - M_0 \sim \left(\frac{M_0}{N - 2}\right) = \triangle M, \text{say} \qquad (12.25)$$

Now let us vary the stripping coefficients T(h) and γ(h) with respect to their estimated values $\hat{T}$(h) and $\hat{\gamma}$(h) and express ΔM in terms of these variations.
    Let

$$x = T(h) - \hat{T}(h))$$

and

$$y = \gamma(h) - \hat{\gamma}(h)$$

then we can write

$$M - M_0 = \triangle M = \sum_{i=0}^{n} \{(xTh_{Gi} - yU_{Gi})\}^2 \qquad (12.26)$$

Using earlier notations (12.17), we can rewrite (12.26) as follows

$$\triangle M = x^2 S_{11} + y^2 S_{22} - 2xy S_{12} \qquad (12.27)$$

For fixed values of ΔM and S coefficients, this Equation (12.27) represents an ellipse. This equation will lead to the required confidence limits for the stripping coefficients. For finding the confidence limits (CL) of $T(h)$, we apply $\partial \Delta M / \partial y = 0$ condition and solve the Equation (12.27), to finally get.

$$x^2 = \frac{M_0}{N-2} \frac{1}{S_{11}(1-r^2)} \qquad (12.28)$$

or we can write

$$x = \pm \left\{ \frac{M_0^m}{N-2} \frac{1}{(S_{11}(1-r^2))} \right\}^{\frac{1}{2}} \qquad (12.29)$$

Where

$r^2 = \frac{S_{12}^2}{S_{11}.S_{22}}$ and M_0^m is for maximum value of M.

Similarly, the confidence limits of $\gamma(h)$ are given below

$$y = \pm \left\{ \frac{M_0^{max}}{N-2} \frac{1}{S_{11}(1-r^2)} \right\}^{1/2} \qquad (12.30)$$

From the expressions (12.29) and (12.30) we can draw one conclusion about the factors which influence the CL limits of the stripping factors: (1) CL decreases as the square root of the number of degrees of freedom, (2) CL increases as r approaches to 1 (or $S_{12}^2 \to S_{11} S_{12}$).

In a similar manner, the confidence limits of uranium stripping factor $\alpha(h)$ can also be found.

12.5 Effects of Airborne Bi[24] and Error Bias due to It

When using Equations (12.7a) and (12.7b) the inputs were assumed to be airborne radon free. This is not always true. It is of interest to study what effects this radon can have in certain worst situations when inversion layers occur during the airborne surveys. These layers are formed due to temperature inversions and may be produced by warm or cold fronts. This layer inhibits the natural upward movement of the air and air pollutants and the aerosols get trapped in this inversion layer. It does not allow the dispersion of these radon associated aerosols and causes extra counts U_R which get added up in the gross counts U_G. These are monitored by the upward looking detector NaI(Tl) crystals. therefore, let us consider the example of uranium stripping factor $\alpha^b(h)$, say when this airborne radon bias is present. Let us put (12.7a) as follows now:

$$U_G + U_R = \alpha^b (h) Th_G + A \tag{12.31}$$

Then proceeding we can write the optimal solution to $\alpha^b(h)$ as

$$\alpha^B(h) = \frac{S_{12} - S_{10} + S_{B1}}{S_{11}} \tag{12.32}$$

where $S_{B1} = \Sigma U_b Th_G$
This can further be rewritten as

$$\alpha^B(h) = ((S_{12} - S_{10})/S_{11}) + S_{B1}/S_{11}$$

and therefore, the bias due to radon is

$$\triangle \alpha^B(h) = \triangle S_{12}/S_{11} \tag{12.33}$$

Thus, as the airborne radon increases, the bias to $\alpha(h)$ increases always positively but if $S_{11} >> S_{B1}$; this effect reduces. It is possible therefore in highly thoriferous areas that this bias be reduced by virtue of high thorium.

12.6 Ground and Airborne Experiments

In this section we describe some examples of test areas for the application. Two test areas viz (A) and (B) have been selected where ground concentration of U, Th, K are determined by collecting samples at 100 m grid interval. These samples were analyzed by both in-situ (by shielded probe spectrometric) measurements and by laboratory radiometric methods. The number of samples in area A and B were respectively 188 and 117 and the results are summarized in Table 12.1. The comparative summary of measurements in test area A is given in Table 12.2.

It was found that the variation of uranium and potassium is fairly constant along the length of traverse and a test strip of this nature is suitable for the estimation of stripping factors.

The airborne data for U, Th, K were collected by flying over the test strips (where ground concentration of radioelements is reliable known) at every second at different altitudes ranging from approximately 30 to 250 m.

Table 12.1 Details of data and radiometric concentrations two areas

Area	No. of Ground Sample	% K	ppm eU	ppm eTh
A	188	41.0 ± 11.1	2.5 ± 0.2	54.5 ± 12.5
B	117	$1.45 \pm .42$	3.77 ± 1.06	15.63 ± 3.69

Table 12.2 Some comparison of laboratory and in-situ measurements

Laboratory - Assayed Grade			In-Situ Grades		
% K	ppm eU	ppm eTh	% K	ppm eU	ppm eTh
3.9	6	50	3.0	4	35
4.5	2	47	3.9	4	68
4.3	1	38	3.1	2	34
4.5	4	93	4.2	2	88

The temperature and pressure data were also collected. Temperature profile at various heights were also taken which helps to know whether an inversion layer at any flying interval is present or not.

Using the relationship between altitude, temperature and pressure, the measured altitude was reduced to standard temperature and pressure conditions. A typical example of the variation of uranium, thorium, potassium, altitude, temperature, data at planned survey altitude of 120 m.

Two more areas were also selected where the thorium to uranium and thorium to potassium ration was significantly high. Another area was selected far away from this, where uranium was significantly higher than thorium and potassium. Since these areas were flown only at two heights, the stripping factor was only computed at these two altitudes.

12.7 Results and Discussion

Several test areas described in previous section are utilized for determination of the stripping coefficient and their variation. The stripping factors α, β, γ, T were estimated at various altitudes. Some results are given in Tables 12.3 and 12.4.

Table 12.3 Results from studies of OLS and WLS models

Altitude (m)	Standard Error (+)	Stripping Factor OLS	Error (+)	WLS	Error (+)	Constant At Height (h)	Error (+)
35.43	3.14	0.337	0.031	0.350	0.025	136.65	16.60
61.18	2.42	0.370	0.029	0.411	0.034	111.38	17.79
92.27	2.89	0.390	0.037	0.399	0.023	121.01	8.22
121.98	2.07	0.430	0.056	0.435	0.060	117.97	21.22
152.29	3.21	0.440	0.045	0.475	0.040	94.76	10.55
183.48	2.84	0.450	0.062	0.528	0.081	87.74	18.93
215.29	3.80	0.470	0.045	0.501	0.059	98.20	10.49
245.19	5.35	0.536	0.078	0.674	0.080	78.98	11.57

Extrapolated value at ground from WLS model = 0.312 ± 0.025
Extrapolated value at ground from OLS model = 0.316 ± 0.027
The estimated value at Calibration pads = 0.327 ± 0.0278

Table 12.4 Comparative study of stripping factors

Model	Thorium into Uranium			Stripping Ratio Increase		Crystal Dimensions (mm)
	0 m	50 m	125 m	0 m-50 m	0 m-125 m	
OLS	0.33	0.36	0.41	0.03	0.080	$406 \times 102 \times 102$
WLS	0.33	0.38	0.45	0.05	0.125	$406 \times 102 \times 102$
Lovborg	0.37	0.40	0.43	0.037	0.060	292×102

12.7.1 Test Area A

In this area average uranium is 2.5 ppm and comparatively lower than thorium presented in Table 12.1. The theoretically computed values by Lovborg et al. (1978), Grasty (1975), IAEA (1991) factors are also considered for comparison.

12.7.2 Test Area B

The particulars of this area are provided in Table 12.1 where the ratios U/Th and Th/K are much less as compared to those of test area A. The variation of uranium and potassium is also not as less as in case of test area A. Therefore, it presents the estimation of stripping factors under a different condition.

12.7.3 Area of Very High Thorium Concentration

Some airborne profiles of the area where the computed concentration of radioelements using calibration data are recorded as U = 1 ppm, Th = 250 ppm and K = 2%. The values of α, β, γ, were estimated at 120 m and 150 m.

12.7.4 Area of High Uranium Concentration

Some areas were selected where average uranium was on higher side and then this area was mainly used to determine the γ coefficient at 120 m, and 150 m heights.

12.8 Concluding Remarks

The variation of the stripping factors for various energies of the gamma rays in airborne surveys are studied from the point of view of direct measurements for radiometric counts in different energy windows and are given in Table 12.4 for comparison purpose. The main concern has been the uranium, thorium, potassium radioelements whose gamma energies

1.76 MeV (Bi-214), 2.62 MeV (Tl-208) and 1.46 MeV (K-40) are considered and their Compton scattering and consequent stripping in lower energy windows. The estimation of these stripping factors and their variation with flying altitude, h, is considered with uranium, thorium, potassium rich areas and the error ellipse has been considered to study their confidence levels. It is interesting to note that using WLS method the results are more dependable compared to OLS model. The data has been compared for Compton stripping factor for thorium to uranium in these results to that estimated at ground level for calibration facilities.

References

[1] Dickson, B.H., Bailey, R.C., and Grasty, R.L. 1981. Utilizing multichannel airborne gamma ray spectrometric, Can. Jour. Earth Sci. v. 18, pp. 1792–1801.

[2] Grasty R.L., 1975, Uranium measurement by airborne gamma ray spectrometry, Geophysics, v. 40, pp. 503–519.

[3] International Atomic Energy Agency, 1991 – airborne Gamma ray spectrometer surveying Technical reports series No. 323 – IAEA Vienna, Australia.

[4] Jadhav, J.G. Atal, B.S. Yudhishtar Lall and Surange, P.G. 1988 – Development as a facility for calibration of airborne and portable gamma ray spectrometers, Exploration & Res. Atomic Minerals vol. 1, pp. 122–128.

[5] Loveborg, L., Grasty, R.L. and Kirkegaard P., 1978, A guide to the calibration constants for aerial gamma-ray surveys in geoexploration: A, Nucl. Sol. on Aerial Techniques for Environmental Monitoring, 193, 206.

[6] Loveborg, L., 1984 – The calibration as portable and airborne gamma ray spectrometer theory problems and facilities. Report No. Riso-M-2456, Nat. Lab, Denmark.

[7] Raghuwanshi, S.S. (1992). Airborne gamma-ray spectrometry in uranium Exploration, Adv. Space Res., v. 12, n. 7, pp. 77–86.

[8] Raghuwanshi, S.S., Bhaumik, B.K. (1989). Estimation of stripping ratios at survey height in airborne gamma ray spectrometry. Nucl. Geophysics, v. 3, n. 2, pp. 113–115.

[9] Raghuwanshi, S.S., Bhaumik, B.K. and Tewari, S.G., (1989). A direct method for determining the altitude variation of the uranium-stripping ratio in airborne gamma ray surveys. Geo-physics, v. 54, n. 10, pp. 1350–1353.

[10] Raghuwanshi, S.S. and Tewari, S.G (1990). Implementation of the inverse distance weighing technique for airborne radiometric data, Nucl. Geophysics, v. 4, n. 2, pp. 259–270.

[11] Tewari, S.G., and Raghuwanshi, S.S. (1987). Some problems on the range of investigation in airborne gamma-ray spectrometry. Uranium, v. 4, pp. 67–82.

13

Modeling Kernel Function in Blackbody Radiation Inversion

Shivvaran Singh Raghuwanshi[1,2]

[1]Ex-Head Physics Group, Atomic Minerals Directorate for Exploration & Research, Department of Atomic Energy, Hyderabad, India
[2]Bhopal (M.P.), India
E-mail: raghu2257@yahoo.com

The inverting of the area temperature profile a(T) from the measured total power spectrum $W(\nu)$ of temperature at different frequencies has been challenging. And the inversion solution of this blackbody radiation problem has been dealt using Planck radiation law. A kernel function comes into these solutions and this mathematical kernel function plays a pivotal role in the inversion processing. It has been modeled in the present treatment to find out the feasible solution for the temperature distribution function by deconvolving the governing equation using the Faltung theorem. The numerical modeling of this kernel function and subsequently their Fourier transforms then is applied to extract the temperature distribution. The results have been pictorially presented in this paper and have been found satisfactory. This approach shall be faster and simpler compared to the earlier works to handle the temperature problem particularly and may be found useful in remote sensing. Rigorous attention is paid to this problem based on a few inversion approximations for easy numerical techniques and is presented in figures for better visual accuracy, comparison, and efficiency.

13.1 Introduction

In physical and allied scientific fields, it is interesting to make inferences about the functions or parameters that are the main cause of the observed effects and need to be inverted in the deconvolution sense. This can only be done through the systematic handling of the observed data and taking care of noise associated with them. In general, the laws of physics provide proper guidance to reach the necessary parameters with suitable mathematical transformations and help to compute the data values using an appropriate model.

In the forward problem, the aim is to compute the output data using the numerical inputs and the variations of the controlling input parameters. On the other hand, in the inverse problem, the aim is to derive input parameters from the observed output data. In the inverse problem, the aim is to reconstruct the governing law or model from a set of measurements. In theory, however, sometimes exactly invertible or solvable situations exist. But quite often the problem of inversion poses severe limitations and hurdles and exact solvability does not occur. In such a case the combination of both theoretical, as well as optimal numerical data processing using Fourier or Laplace transforms are helpful in dealing with these kinds of problems.

Blackbody radiation is also such a problem where the power density at a particular frequency is governed by the Planck law and here temperature and frequency play the central role. But when we observe the power density for all frequencies, extraction of the temperature distribution function from this data becomes a very complex problem and the problem now needs special attention to be solved. The power density integrated over a certain range of temperature for a particular frequency takes the form of an integral equation. This is because the temperature distribution function that needs to be unfolded remains in the integrand.

This blackbody radiation problem has been dealt in the past by many workers [1–11] in various ways and challenges are always there to achieve a better solution to the problem. Here in this work, we have tried to design another approach to solve this equation by studying the behavior and then modeling the kernel function and may be found interesting.

13.2 Nature and Statement of the Problem

To find the temperature distribution a(T) from the measured power spectrum $W(\nu)$ of a blackbody having temperature T and emitting a radiation of frequency ν is an interesting problem. This is an inversion problem in which we need to extract the information about the temperature distribution with the prior knowledge of the other parameters such as frequencies of radiation. Historically, Wien had attempted to relate the frequency and temperature of the blackbody rations before Planck had reached to the ultimate radiation law. He derived it from the logic of thermal equilibrium and was able to interpret the blackbody for large frequencies while Rayleigh was able to interpret the same for short frequencies of this complex problem. But Planck took the quantization of radiation and assumed all photons as simple harmonic oscillators and derived the law (13.1) radiation law.

The inversion problem we are dealing in this work relates to the Planck's law of radiation. In Planck's law, these three quantities viz. P, ν, T are well related via the energy of photons which are harmonic oscillators. The power spectrum $P(\nu, T)$ with absolute temperature T of the radiating body is expressed as follows

$$P(\nu, T) = \frac{\frac{2h\nu^3}{c^2}}{e^{\frac{h\nu}{kT}} - 1} \tag{13.1}$$

where,

h is Planck constant,
c is the speed of light,
k is Boltzmann constant,
T is the temperature of the blackbody.

Wein had found this law for large frequencies or short-wave lengths of radiation before the Equation (13.1) came into being. The amount of energy, $Pw(\nu, T)$ per unit surface area per unit time per unit solid angle per unit frequency emitted at a frequency ν according to Wein's law can be written for higher frequencies or short wavelengths in the following manner

$$P(\nu, T) = \left(\frac{2h\nu^3}{c^2}\right) e^{-\frac{h\nu}{kT}} \tag{13.2}$$

Obviously, from Equations (13.1) and (13.2) of power densities, we can see that the maxima of these two densities two do not coincide. The maximum P predicted by Planck law is higher than that computed by Wein law because of $[e^{h\nu/kT} - 1] < e^{h\nu/kT}$. This is also important to see that for frequencies when $h\nu \gg kT$, the expression (13.1) for the density $P(\nu, T)$ reduces to the expression (13.1) that derived by Wein. In fact, this law will be very useful for the inversion of the temperature at higher frequencies or short-wave lengths.

When we take the logarithm of both sides of Equation (13.1), we shall get the following expression of the temperature

$$T = \frac{h\nu}{k \log\left(\frac{2h\nu^3}{Pc^2}\right)} \tag{13.3}$$

The expression of temperature (13.2) allows us to compute the temperature for different values of power densities P and frequencies such that $h\nu \gg kT$.

For all temperatures, if we measure the spectral density $W(\nu, T)$ then the area distribution of temperature will not be such a simple formula but turn into an integral for a range of temperatures. In this context, it is interesting to note that the Wein approximation (13.3) for describing the radiation regime of radiation is very useful for higher frequencies while for a whole range of the radiation frequencies from low to higher the Planck formula (13.1) shall be useful.

When the effective area temperature distribution function a(T) as a function of temperature is integrated over all possible temperatures, the total power spectrum radiated by this blackbody is expressed as follows

$$W(\nu) = \left(\frac{2h\nu^3}{c^2}\right) \int_0^\infty a(T)dT/(e^{h\nu/kT}-1) \qquad (13.4)$$

For $h\nu \gg kT$ the above formula (13.4) will reduce to the following equation:

$$W(\nu) = \left(\frac{2h\nu^3}{c^2}\right) \int_0^\infty e^{-h\nu/kT} a(T)dT \qquad (13.5)$$

The Equations (13.4) and (13.5) are the integral equations for the unknown function a(T) to compute the temperature profile from the measured quantities $W(\nu)$ at different known frequencies.

In the case of long wavelengths of radiation, Rayleigh-Jeans had found that the energy output goes to infinity as the frequencies approach to zero and this was its failure to explain the energy output in accordance with observations. This case was easily derivable from the Equation (13.5) for $h\nu \ll kT$ and this condition will render the following equation for $P(\nu, T)$ and $W(\nu)$ in this case

$$P(\nu, T) = \left(\frac{2h\nu^3}{c^2}\right)\left(\frac{kT}{h\nu}\right) \qquad (13.6)$$

$$W(\nu) = \left(\frac{2h\nu^3}{c^2}\right) \int_0^\infty \left(\frac{kT}{h\nu}\right) a(T)dT \qquad (13.7)$$

All these three Equations (13.4), (13.5) and (13.6) are the integral equations for unknown temperature function a(T). The solution of such an integral equation is very interesting and poses a challenge for solving it for the extraction of the temperature distribution function; it is tackled by the deconvolution methods using suitable transformations of the variables. In practice, there are many ways in which this can solve but mostly the common approach is to compress the variables by exponential transformations which have been the practice in linear filter theory. This has the advantage that the resulting transformed integral equation can be treated by the method of FFT and the deconvolution process converts it into a simple algebraic case from which unknown can be determined. This transformation, however, generates a shaping or spread function which is called a kernel function whose Fourier transform plays the central role in determining the accurate results. That encourages us to study this kernel function in detail and simpler to use for extraction of the temperature information. The nature of kernel functions is in itself very interesting from mathematical points of views too. The behavior of this kernel function entirely depends on the transformations one chooses to reduce the integral equation to the convolution form. The small variations in these transformations can give rise to very useful visualizations. This can help us sometimes to compute simple filters for deconvolutions.

13.3 Towards a Solution to the Problem

The integral so obtained must be put in such a form that is amenable for the application of convolution theorems. The advantage with these theorems is that they treat the overlapping of one function over the other in a multiplicative way and help us to visualize the hidden nature of real causative function with the help of the kernel functions. The shape and simplicity of these kernel functions are most important in these treatments and are the results of proper transformations. Mostly the exponential transformations are very suitable for reducing the integral equations to convolution forms and yield wonderful results in deconvolving the causative function from the observed data. However, this is not a common method for all integral equations. This linear filter theory becomes very useful for the development of recursive filters also. A set of coefficients which are computed by processing the resulting equations by FFT are used to provide acceptable filters with proper choice of S/N ratios.

Chen (1990) treated this problem by converting this integral Equation (13.6) into a convolution equation by adopting the following transformations in the variables

$$e^x = \frac{h\nu}{kT_0}$$

$$e^y = \frac{T}{T_0} \tag{13.8}$$

Assuming T_0 as some reference temperature for convenience and to make the transformations dimensionless. Thus, the frequency ν is expanded exponentially with increasing x but x remains small for large frequencies.

On substituting and differentiating these new variables in the appropriate places in the integral Equation (13.7) gives a new following convolution formula

$$G(x) = \int_{-\infty}^{+\infty} \phi(x-y)A(y)dy \tag{13.9}$$

where the various new symbols are as below:

$$G(x) = \left(\frac{h^2c^2}{2k^3T_0^4}\right) W\left(kT_0e^x/h\right) e^{-x(2-\Delta)}$$

$$A(y) = a(T_0e^y)e^{-y(2-\Delta)} \tag{13.10}$$

Here the $\Phi(x-y)$ is the kernel function which only depends on the variable $(x-y)$ which is a dimensionless function. It has a specific shape defined by the following mathematical expression obtained consequently after the substitutions.

$$\Phi(x-y) = \frac{e^{(1+\Delta)(x-y)}}{e^{e^{-y}}-1} \tag{13.11}$$

Let us assume $u = x-y$ and rewrite this kernel function again as follows:

$$\Phi(u) = e^{(1+\Delta)u}/(e^{e^u}-1) \tag{13.12}$$

Here an arbitrary numerical adjustable factor Δ was introduced by Chen (1990) which helps to decide the width of this kernel function. This kernel function is negatively skewed function while for deconvolution a symmetrical kernel is always preferred for the deconvolving operations. Chen has graphically shown this nature by varying Δ from 0 to 3.

The kernel functions for Wein and Rayleigh-Jeans laws too can similarly be derived from this shape function (13.12). They will look like as given below:

For Wein

$$\Phi(u) = e^{(1+\Delta)u}/(e^{e^u})$$ (13.13)

And for Rayleigh-Jeans

$$\Phi(u) = \frac{e^{(1+\Delta)u}}{e^u}$$ (13.14)

But from all these asymmetrical kernel functions we can break them up in near symmetrical kernel functions. In the present work, this has been attempted and the dummy parameter delta has been varied accordingly. And it has been used to model such that the kernel $\Phi(x)$ can be brought nearer to a Gaussian function or combination of Gaussians. These Gaussian functions are therefore symmetrical and can be utilized more smoothly than the cases of kernel functions defined in (13.2), (13.3) and (13.4). The shape of these three kernel functions for $\Delta = 1$ is given in Figure 13.1. We see that the shapes of these three kernel functions are similar to the radiation law underlying the integration process and it is justified also because we have transformed the Planck radiation law into a convenient convolution format. The role of these functions is very important to extract the behavior of the area distribution of the temperature profile a(T). This is also interesting to note that the Wein law predicts the true energy outputs on higher frequencies and so the corresponding kernel function to matches at higher values of u so well like the law behavior. This is the true reflection of the kernel function behavior of the law. As in the case of Rayleigh-Jeans law, the predictability

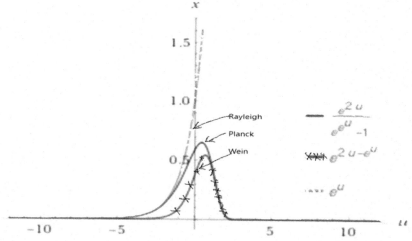

Figure 13.1 Shapes of kernel functions for parameter $\Delta = 1$.

of energy output is true for lower frequencies or long wavelengths, this is also reflected by its kernel function. The kernel function due to Planck radiation law is all through true and is just similar in shape to its radiation curve. But all these curves are skewed in nature. This poses a serious problem in the inversion process of the temperature profile of Equation (13.4). Since we cannot exactly find symmetry in these kernel functions, we can try to achieve a reasonable symmetrical picture by managing the adjustable parameter and tolerance levels for reaching nearer to the symmetry by suitable modeling either heuristically or statistically.

13.3.1 Reducing $\Phi(x)$ Function to Gaussians

The kernel functions as we have seen in the earlier section are skewed and we have to reach to a solution which is not only nearer to its peak value but the width also to a sufficiently satisfactory level. This has been done by computing the basic kernel function by fixing its maximum value as its width is not in our control. Then a Gaussian function with the same peak value is selected. And the width of this Gaussian function is controlled by its spread parameter, called in statistical terms, a standard deviation. This parameter is constantly varied, and it is monitored until it matches or overlaps both the sides of the kernel function under consideration. Well, this matching will not be perfect, but we will accept it if the overlapping is reasonable.

The modelling kernel functions in this paper have been treated recursively and reduced gradually by following this heuristic approach to nearly Gaussian functions as shown in Figures 13.2 and 13.3.

In Figure 13.2, we have taken the dummy parameter $\Delta = 6.5$ and the corresponding kernel function in blue color is modeled for a Gaussian function with a centroid at 2.0143 and standard deviation as 0.36055 and of maxima at 2002.3. The matching is self-explanatory in this diagram.

As the kernel function (13.4) is always skewed negatively, we have also attempted to break it up using error function also. The error function is originally is related to the Gaussian function of probability and can be thought similar to the portion of this kernel function on the left-hand side of the origin. From the various families of standard curves, we have found this as most suitable for our work and there are no special criteria for its selection.

Here we have taken $\Delta = 2$ and plotted it in Figure 13.3 and shown in blue color. Here best fit Gaussian and erf(x) function was attempted. This is plotted on this kernel function itself as shown in this Figure 13.3 in red color.

Here too the matching is found to be satisfactory.

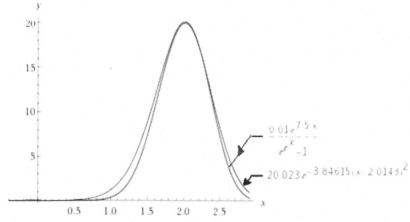

Figure 13.2 Modeled gaussian kernel superimposed on actual one for parameter $\Delta = 6.5$.

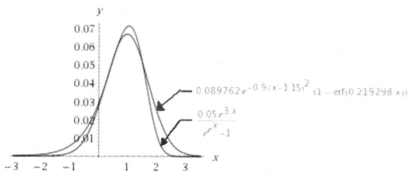

Figure 13.3 Modeled error function kernel function superimposed on actual one for parameter $\Delta = 3.0$.

The mathematical expressions for kernel function are given in the following two instances as follows:

Kernel function (13.2) of this radiation problem is given as follows for dummy parameter $\Delta = 6.5$:

$$\frac{e^{7.5x}}{-1 + e^e} \tag{13.15}$$

Let us denote the modeled kernel function in this case by the function $\Phi 1(x)$ as follows:

$$\Phi_1(x) = 2002.3e^{-3.84615(-2.0143+x)^2} \tag{13.16}$$

In a similar way we can take a parameter, $\Delta = 2$ in the kernel (13.1) for computing another modeled kernel $\Phi2(x)$, say in this case as given below:

$$\frac{e^{3x}}{-1 + e^e} \tag{13.17}$$

$$\Phi_2(x) = 0.089762e^{-1.8(-1.15+x)}(1 - \text{erf}(0.219298\,x)) \tag{13.18}$$

The behavior of these kernel functions and their modeled functions are plotted in Figures 13.2 and 13.3 respectively. The deviations of these two models (13.10) and (13.12) from their original parent kernel functions (13.9) and (13.11) are just nominal and are acceptable in 90% limit of good fits.

13.3.2 Convolution of Equation (13.4)

The convolution integral Equation (13.4) now can be rewritten as follows with these newly modeled kernel functions (13.10) and (13.12) as expressed above by appropriate substitutions

$$G(x) = \int_{-\infty}^{\infty} \phi_1(x - y)A(y)dy \tag{13.19}$$

$$G(x) = \int_{-\infty}^{\infty} \phi_2(x - y)A(y)dy \tag{13.20}$$

Thus, these two above equations are in convolution format and are amenable for the application of Fourier transform. By taking Fourier transform (FT) of both the sides of Equation (13.20) and using convolution theorem we can write the following:

$$FT(G(x)) = FT(\phi_1(x - y)) * FT(A(y)) \tag{13.21}$$

which with little algebraic manipulation can be written to find the FT(A(y)). And then taking inverse FT of both the sides of that expression, subsequently we find the following inversion for the required temperature distribution function A(y)

$$A(y) = FT^{-1}\left(\frac{FT(Gx)}{FT(\phi_1(x - y))}\right) \tag{13.22}$$

The Fourier transform of $\Phi1(x-y)$ is given in Figure 13.4. as follows:

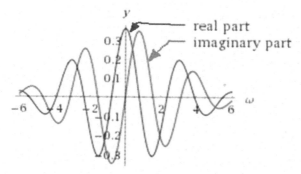

Figure 13.4 Fourier transform of $\Phi 1(x-y)$.

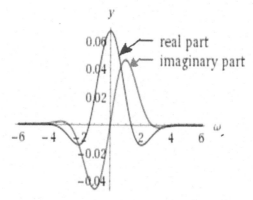

Figure 13.5 Fourier transform of $\Phi 2(x-y)$.

And FT of Gaussian exponential part of $\Phi 2(x-y)$ is shown in Figure 13.5 and mathematically expressed as follows:

$$\mathcal{F}_x \left[20.023 \exp\left(-0.5 \times \frac{(x - 2.0143)^2}{0.13} \right) \right] (w)$$
$$= 1.20548 \times 10^{-6} e^{-0.065(\omega - 15.4946i)^2}$$

And

$$\mathcal{F}_x \left[0.089762 \exp\left(-0.9(x - 1.15)^2 \right) \right] (w)$$
$$= 0.0203487 e^{-0.277778(\omega - 2.07i)^2}$$

Where $\mathcal{F}_x$ is the Fourier Transform of f (x) with frequency ω.

13.4 Concluding Remarks

The heuristic approach adopted in this analysis of breaking up the two kernel functions has simplified by focusing their inherent skewed nature. It simplifies the inversion procedure for the extraction of temperature profile from the observed data, $W(\nu)$ in a better way compared to the direct usage of the kernel functions (i.e. $\frac{e}{-1+e^e}$). The nature of this function is unraveled better by varying the dummy parameter delta. This could help us to study the behavior of this kernel function and we could express it in Gaussian terms. This approach makes the deconvolution treatment of the problem under consideration very simple by directly using its Fourier Transform and it is convenient for dealing with real data too.

References

[1] Bojarski N. Inverse black body radiation. IEEE Trans Antennas Propagation, 1982, 30: 778–780.

[2] Dai X X, Dai J Q. On unique existence theorem and exact solution formula of the inverse black-body radiation problem. IEEE Trans Antennas Propagation, 1992, 40: 257–260.

[3] Hamid M, Ragheb H A. Inverse black body radiation at microwave frequencies. IEEE Trans Antennas Propagation, 1983, 31: 810–812.

[4] Kim Y, Jaggard D L. Inverse black body radiation: An exact closed-form solution. IEEE Trans Antennas Propagation, 1985, 33: 797–800.

[5] Hunter J D. An improved closed-form approximation to the inverse black body radiation problem at microwave frequencies. IEEE Trans Antennas Propagation, 1986, 34: 261–262.

[6] Chen N X. A new method for inverse black body radiation problem. Chin Phys Lett, 1987, 4: 337–340.

[7] Lakhtakia M N, Lakhtakia A. Inverse black body radiation at submillimeter wavelengths. IEEE Trans Antennas Propagation, 1984, 32: 872–873.

[8] Chen N X. Theoretical Investigation on the Inverse Blackbody Radiation. IEEE Trans Antennas Propagation, 1990, 38: 1287–1290.

[9] Dai X X, Dai J Q. Exact solutions of the inverse black-body radiation problem hierarchy with Planck integral spectrum. Phys Lett A, 1991, 161: 45–49.

[10] Lakhtakia M N, Lakhtakia A. On some relations for the inverse black-body radiation problem. Appl Phys B, 1986, 39: 191–193.

[11] Sun X G, Jaggard D L. The inverse blackbody radiation problem: A regularization solution. J Appl Phys, 1987, 62: 4382–4386.

Index

About the Editors

Mangey Ram received the PhD degree major in mathematics and minor in computer science from G. B. Pant University of Agriculture and Technology, Pantnagar, India. He has been a faculty member for around 10 years and has taught several core courses in pure and applied mathematics at undergraduate, postgraduate, and doctorate levels. He is currently a professor at Graphic Era (Deemed to be University), Dehradun, India. Before joining the Graphic Era, he was a deputy manager (Probationary Officer) with Syndicate Bank for a short period. He is Editor-in-Chief of *International Journal of Mathematical, Engineering and Management Sciences* and the Guest Editor & Member of the editorial board of various journals. He is a regular reviewer for international journals, including IEEE, Elsevier, Springer, Emerald, John Wiley, Taylor & Francis, and many other publishers. He has published 150 plus research publications in IEEE, Taylor & Francis, Springer, Elsevier, Emerald, World Scientific, and many other national and international journals of repute and also presented his works at national and international conferences. His fields of research are reliability theory and applied mathematics. Dr. Ram is a Senior Member of the IEEE, life member of Operational Research Society of India, Society for Reliability Engineering, Quality and Operations Management in India, Indian Society of Industrial and Applied Mathematics, member of International Association of Engineers in Hong Kong, and Emerald Literati Network in the U.K. He has been a member of the organizing committee of a number of international and national conferences, seminars, and workshops. He has been conferred with *"Young Scientist Award"* by the Uttarakhand State Council for Science and Technology, Dehradun, in 2009. He has been awarded the *"Best Faculty Award"* in 2011; *"Research Excellence Award"* in 2015; and recently *"Outstanding Researcher Award"* in 2018 for his significant contribution in academics and research at Graphic Era (Deemed to be University, Dehradun, India).

Tadashi Dohi received his Bachelor, Master and Doctorate of Engineering degrees from Hiroshima University, Japan, in 1989, 1991, and 1995, respectively. In 1992, he joined the Department of Industrial and Systems Engineering, Hiroshima University, as an assistant professor. In 1992 and 2000, he was a visiting research fellow in the Faculty of Commerce and Business Administration, University of British Columbia, Canada, and Hudson School of Engineering, Duke University, USA, respectively. Since 2002, he has been working as a full professor in the Department of Information Engineering, Hiroshima University. He is now Vice Dean of School of Informatics and Data Science, Hiroshima University. Dr. Dohi's research interests include reliability engineering, software reliability, dependable computing, performance evaluation, computer security, and operations research. He is a regular member of ORSJ, IEICE, IPSJ, REAJ, IEEE Computer Society, and IEEE Reliability Society. He has published over 550 peer-reviewed articles, 30 book chapters, and 20 edited books/proceedings in the above research areas. Dr. Dohi served as the General Chair of 12 international conferences, including ISSRE 2011 and DASC 2019, and as the Program Committee Chair of three international events. He has worked as a Program Committee Member in many international conferences such as DSN, ISSRE, SRDS, COMPSAC, QRS, EDCC, PRDC, HASE, among many others. He is an associate editor/editorial board member of over a dozen international journals including IEEE Transactions on Reliability. He is now the President of REAJ (Reliability Engineering Association of Japan) and the Executive Committee Member of ORSJ (Operations Research Society of Japan).